4

COLLOQUIA

Yves André
Département de Mathématiques et Applications
École Normale Supérieure
45 rue d'Ulm
75230 Paris Cedex 05, France

Nicholas M. Katz
Department of Mathematics
Princeton University
806 Fine Hall
Princeton, NJ 08544-1000, USA

Endre Szemerédi
Alfréd Rényi Institute of Mathematics
Hungarian Academy of Sciences
Reáltanoda n. 13-15
H-1053 Budapest, Hungarian

Dan Abramovich
Department of Mathematics
Brown University
Box 1917
Providence, RI 02912, USA

Elias M. Stein
Department of Mathematics
Princeton University
802 Fine Hall
Princeton, NJ 08544-1000, USA

Shou-Wu Zhang
Department of Mathematics
Princeton University
505 Fine Hall
Princeton, NJ 08544-1000, USA

Colloquium De Giorgi 2010-2012

edited by
Umberto Zannier

EDIZIONI
DELLA
NORMALE

ISBN 978-88-7642-455-7
e-ISBN 978-88-7642-457-1

Contents

Preface

Since 2001 the Scuola Normale Superiore di Pisa has organized the "Colloquio De Giorgi", a series of colloquium talks named after Ennio De Giorgi, the eminent analyst who was a Professor at the Scuola from 1959 until his death in 1996.

The Colloquio takes place once a month. It is addressed to a general mathematical audience, and especially meant to attract graduate students and advanced undergraduate students. The lectures are intended to be not too technical, in fields of wide interest. They should provide an overview of the general topic, possibly in a historical perspective, together with a description of more recent progress.

The idea of collecting the materials from these lectures and publishing them in annual volumes came out recently, as a recognition of their intrinsic mathematical interest, and also with the aim of preserving memory of these events.

For this purpose, the invited speakers are now asked to contribute with a written exposition of their talk, in the form of a short survey or extended abstract. This series has been continued in a collection that we hope shall be increased in the future.

This volume contains a complete list of the talks held in the "Colloquio De Giorgi" in recent years but also in the past and also in the past years, and a table of contents of all the volumes too.

Colloquia held in 2001

Paul Gauduchon
Weakly self-dual Kähler surfaces

Tristan Rivière
Topological singularities for maps between manifolds

Frédéric Hélein
Integrable systems in differential geometry and Hamiltonian stationary Lagrangian surfaces

Jean-Pierre Demailly
Numerical characterization of the Kähler cone of a compact Kähler manifold

Elias Stein
Discrete analogues in harmonic analysis

John N. Mather
Differentiability of the stable norm in dimension less than or equal to three and of the minimal average action in dimension less than or equal to two

Guy David
About global Mumford-Shah minimizers

Jacob Palis
A global view of dynamics

Alexander Nagel
Fundamental solutions for the Kohn-Laplacian

Alan Huckleberry
Incidence geometry and function theory

Colloquia held in 2002

Michael Cowling
Generalizzazioni di mappe conformi

Felix Otto
The geometry of dissipative evolution equations

Curtis McMullen
Dynamics on complex surfaces

Nicolai Krylov
Some old and new relations between partial differential equations, stochastic partial differential equations, and fine properties of the Wiener process

Tobias H. Colding
Disks that are double spiral staircases

Cédric Villani
When statistical mechanics meets regularity theory: qualitative properties of the Boltzmann equation with long-range interactions

Colloquia held in 2003

John Toland
Bernoulli free boundary problems-progress and open questions

Jean-Michel Morel
The axiomatic method in perception theory and image analysis

Jacques Faraut
Random matrices and infinite dimensional harmonic analysis

Albert Fathi
C^1 *subsolutions of Hamilton-Iacobi Equation*

Hakan Eliasson
Quasi-periodic Schrödinger operators-spectral theory and dynamics

Yakov Pesin
Is chaotic behavior typical among dynamical systems?

Don B. Zagier
Modular forms and their periods

David Elworthy
Functions of finite energy in finite and infinite dimensions

Colloquia held in 2004

Jean-Christophe Yoccoz
Hyperbolicity for products of 2×2 *matrices*

Giovanni Jona-Lasinio
Probabilità e meccanica statistica

John H. Hubbard
Thurston's theorem on combinatorics of rational functions and its generalization to exponentials

Marcelo Viana
Equilibrium states

Boris Rozovsky
Stochastic Navier-Stokes equations for turbulent flows

Marc Rosso
Braids and shuffles

Michael Christ
The d-bar Neumann problem, magnetic Schrödinger operators, and the Aharonov-Böhm phenomenon

Colloquia held in 2005

Louis Nirenberg
One thing leads to another

Viviane Baladi
Dynamical zeta functions and anisotropic Sobolev and Hölder spaces

Giorgio Velo
Scattering non lineare

Gerd Faltings
Diophantine equations

Martin Nowak
Evolution of cooperation

Peter Swinnerton-Dyer
Counting rational points: Manin's conjecture

François Golse
The Navier-Stokes limit of the Boltzmann equation

Joseph J. Kohn
Existence and hypoellipticity with loss of derivatives

Dorian Goldfeld
On Gauss' class number problem

Colloquia held in 2006

Yuri Bilu
Diophantine equations with separated variables

Corrado De Concini
Algebre con tracce e rappresentazioni di gruppi quantici

Zeev Rudnick
Eigenvalue statistics and lattice points

Lucien Szpiro
Algebraic Dynamics

Simon Gindikin
Harmonic analysis on complex semisimple groups and symmetric spaces from point of view of complex analysis

David Masser
From 2 to polarizations on abelian varieties

Colloquia held in 2007

Klas Diederich
Real and complex analytic structures

Stanislav Smirnov
Towards conformal invariance of 2D lattice models

Roger Heath-Brown
Zeros of forms in many variables

Vladimir Sverak
PDE aspects of the Navier-Stokes equations

Christopher Hacon
The canonical ring is finitely generated

John Coates
Elliptic curves and Iwasava theory

Colloquia held in 2008

Claudio Procesi
Funzioni di partizione e box-spline

Pascal Auscher
Recent development on boundary value problems via Kato square root estimates

Hendrik W. Lenstra
Standard models for finite fields

Jean-Michel Bony
Generalized Fourier integral operators and evolution equations

Shreeram S. Abhyankar
The Jacobian conjecture

Fedor Bogomolov
Algebraic varieties over small fields

Louis Nirenberg
On the Dirichlet problem for some fully nonlinear second order elliptic equations

Colloquia held in 2009

Michael G. Cowling
Isomorphisms of the Figa-TalamancaHerz algebras Ap(G) for connected Lie groups G

Joseph A. Wolf
Classical analysis and nilpotent Lie groups

Gisbert Wustholz
Leibniz conjecture, periods & motives

David Mumford
The geometry and curvature of shape spaces

Colloquia held in 2010

Charles Fefferman
Extension of functions and interpolation of data

Colloquia held in 2011

Ivar Ekeland
An inverse function theorem in C^∞

Pierre Cartier
Numbers and symmetries: the 200th anniversary of Galois' birth

Yves André
Galois theory beyond algebraic numbers and algebraic functions

Colloquia held in 2012

Nicholas M. Katz
Simple things we don't know

Endre Szemeredi
Is laziness paying off? ("Absorbing" method)

Dan Abramovich
Moduli of algebraic and tropical curves

Elias M. Stein
Three projection operators in complex analysis

Shou-Wu Zhang
Congruent numbers and Heegner points

Contents of previous volumes:

Galois theory beyond algebraic numbers and algebraic functions

Yves André

Abstract. Classical applications of Galois theory concern algebraic numbers and algebraic functions. Still, the night before his duel, Galois wrote that his last mathematical thoughts had been directed toward applying his "theory of ambiguity to transcendental functions and transcendental quantities". In this talk, we will outline some more or less recent ideas and results in this direction.

2011 was the bicentenary of Evariste Galois's birth.

Galois's insights lie at the origin of the fundamental notion of group - which Grothendieck, in his "Recoltes et Semailles", counts as one of the two main mathematical innovations of all times, alongside with the invention of zero[1].

Galois was also the first to formulate the principle of correspondence between symmetries and invariants, whose fruitfulness exceeded by far the original context of algebraic equations for which he devised his "ambiguity theory".

The main applications of "ambiguity theory" concern algebraic numbers and algebraic functions. Half a century after Galois, Klein's work "blending Galois with Riemann" started a tradition which, through Poincaré's views about uniformization of curves and fundamental groups, eventually led to Grothendieck's fusion between Galois theory and the theory of coverings, a cornerstone of "arithmetic geometry".

But Galois also foresaw applications of "ambiguity theory" beyond the domains of algebraic numbers and algebraic functions, as evidenced by the end of his testament letter (1832):

"Mes principales méditations depuis quelque temps étaient dirigées sur l'application à l'analyse transcendante de la théorie de l'ambiguïté.

---

[1] Imported in Europa by Leonardo da Pisa, aka Fibonacci!

Il s'agissait de voir *a priori* dans une relation entre quantités ou fonctions transcendantes quels échanges on pouvait faire, quelles quantités on pouvait substituer aux quantités données sans que la relation pût cesser d'avoir lieu. Cela fait reconnaître tout de suite l'impossibilité de beaucoup d'expressions que l'on pourrait chercher.

Mais je n'ai pas le temps et mes idées ne sont pas encore bien développées sur ce terrain qui est immense..."

In this talk, we outline some more or less recent ideas and results which extend "ambiguity theory" to the domains of transcendental functions and transcendental numbers[2].

1. Trancendental functions and Galois groups

1.1. Liouville unearthed Galois's papers and was also the first to pursue Galois's intuitions in the direction of transcendental functions. Instead of asking when an algebraic equation is solvable by radicals, he asked when a linear differential equation is solvable by quadratures (integrals and exponential of integrals).

In 1883, Picard (inspired by Lie's ideas) introduced the *differential Galois group*: the group formed by those automorphisms of the extension of the base function field generated by the solutions and their derivatives, which commute to differentiation.

From the fact that solutions of a linear differential equation form a finite-dimensional vector space instead of a finite set, the differential Galois group is no longer a finite group in general, but a linear algebraic group (matrix group). This is in fact one of sources of the theory of algebraic groups, in the work of Kolchin.

The maximal number of algebraically independent elements (over the base function field) among the solutions and their derivatives is the dimension of the differential Galois group.

One has a differential Galois correspondence which relates differential field extensions and some transformation groups of the spaces of solutions.

1.2. The theory matured slowly. The classification of Galois ambiguities in the context of local analytic differential equations (*i.e.* in the case

---

[2] For more detailed treatments of these topics, see for instance our texts "Ambiguity theory, old and new", Boll. Mat. UMI 6 (2009), and "Idées galoisiennes", Journées X-UPS 2011, http://www.math.polytechnique.fr/xups/vol11.html .

where the base function field is the field of germs of meromorphic functions) is due to J. P. Ramis: there are of three types of Galois ambiguities, which "generate" (*i.e.* are Zariski-dense in) the differential Galois group[3].

1) *Monodromy*: this ambiguity, studied by Riemann in the case of the hypergeometric differential equation, comes from the fact that a solution may not take its initial value when one performs a loop around the singularity.

2) *Exponential scaling*: let us consider the differential equation

$$xy' + y = 0$$

around 0; a solution is $y = e^{1/x}$, and any other solution is a constant multiple of y. The differential Galois group of this equation is the group $\mathbb{C}^{\times}$ generated by these scalings.

3) *Stokes phenomenon*: let us consider the inhomogeneous differential equation

$$xy' + y = x$$

around 0, which Euler already encountered in his famous memoir on divergent series[4], as the equation satisfied by the formal series $\hat{y} = \sum (-1)^n n! x^{n+1}$. Euler used this equation in order to "sum" this divergent series, by identifying it with the "true solution" $y = \int_0^{\infty} \frac{e^{-t/x}}{1+t} dt$, of which $\hat{y}$ is the asymptotic expansion (Euler writes "evolutio") in the plane deprived from the negative real half-line.

In another sector, the asymptotic expansion of y may change, giving rise to the Stokes ambiguities (Stokes discovered them in the middle of the XIX century in the case of Airy equation $y'' = xy$, after having remarked that it was much more efficient to compute the zeroes of the Airy solution by using the divergent asymptotic expansion at ∞ rather than the convergent Taylor expansion at 0 as did Airy[5].

1.3.   The early prospects of differential Galois theory, especially in Drach's plans, were much more ambitious and concerned non-linear differential systems. But in spite of Vessiots's lifelong efforts and of the

---

[3] See *e.g.* M. van der Put, M. Singer, Galois theory of linear differential equations, Springer Grundlehren der Math. Wiss. 328, 2003.

[4] De seriebus divergentibus, published in 1760 at the Académiy of St. Petersbourg. See also http://www.maa.org/news/howeulerdidit.html, june 2006.

[5] In his computation of the position of the replica of a rainbow, a phenomenon put forward by the proponents of the wave-theory of light.

relative successes of differential algebra (Ritt, ...), it soon appeared that the foundations of differential Galois theory were not solid enough beyond the linear case.

After various attempts by Pommaret and Umemura, Malgrange's theory seems to provide the right setting in the non-linear case[6]. Differential Galois groups are replaced by "algebraic D-groupoids", defined by algebraic systems of partial differential equations, using jets.

2. Trancendental numbers and Galois groups

2.1. Can one attach to a given transcendental number its conjugates and a Galois group which permutes them transitively?

If so, one might expect the number of conjugates and the Galois group to be infinite... Let us make a first test with π (which Lindemann proved to be transcendental in 1882). As Euler observed, π satisfies an equation of "infinite degree" with rational coefficients, namely[7]

$$\prod_{n\in\mathbb{Z}\setminus 0} \left(1 - \frac{x}{n\pi}\right) = \frac{\sin x}{x} = 1 - \frac{x^2}{6} + \frac{x^4}{120} + \cdots \in \mathbb{Q}[[x]]$$

which suggests to consider the multiple of π as conjugates. In fact, if one wants a group to act on the conjugates, one has to allow any non-zero rational multiple of π, and the corresponding Galois group would be $\mathbb{Q}^\times$.

Can one generalize this approach? An old result by Hurwitz states that any complex number α is a zero of a formal power series f with rational coefficients which converges everywhere, which suggests to considers the other zeroes as the conjugates of α. Unfortunately, this is a dead-end, since there are uncountably many such f's, and no way to choose a canonical one in general.

2.2. We shall see that one can expect, nevertheless, to be able to define conjugates and Galois groups for a vast class of numbers, including most classical mathematical constants. These numbers are integrals, the so-called *periods*[8] $\int_\Delta \omega$, and more generally the *exponential periods* $\int_\Delta e^f \omega$

---

[6] B. Malgrange "On nonlinear differential Galois theory". Frontiers in mathematical analysis and numerical methods, 185–196, World Sci. Publ., 2004.

[7] Formally, Euler's identity $\sum n^{-2} = \pi^2/6$ results from the extension to this equation of Newton's formula for the sum of squares of thes roots in terms of the coefficients. Incidentally, Galois pointed out explicitly that Newton-like formulas do not depend on the degree of the equation; he may have had in view the possibility of letting the degree tend to infinity.

[8] The name comes from the one-dimensional case, where these numbers are periods (in the usual sense) of the corresponding abelian functions.

(where both ω and the domain Δ are defined by algebraic expressions of one or several variables with algebraic coefficients: for instance $\pi = \int_0^\infty \frac{2dt}{1+t^2}$.

The idea comes from Grothendiek's programmatic theory of motives, which aims at unifying combinatorial, topological and arithmetical aspects of algebraic geometry. These motives play somehow the role of "elementary particles" of algebraic geometry, liable to be decomposed and recombined according to representation-theoretic rules. The relevant groups are the so-called "motivic Galois groups", which are generalizations of usual Galois groups to systems of algebraic equations in several variables. As for differential Galois groups, these are not finite groups in general but algebraic groups.

As symmetry groups of motives (defined over $\overline{\mathbb{Q}}$, say), these motivic Galois groups are expected to act on their periods. The coherence of such an action requires to postulate that any algebraic relation (with algebraic coefficients) between periods should come from a relation between motives: this is *Grothendieck's period conjecture*.

It would follow that the maximal number of algebraically independent periods of a motive defined over $\overline{\mathbb{Q}}$ is the dimension of the associated motivic Galois group[9].

For instance, $2\pi i = \int dt/t$ is the period attached to the motive of the line deprived from the origin, whose motivic Galois group is $\mathbb{Q}^\times$. The conjugates of $2\pi i$ are its non-zero rational multiples, and the coherence postulated by Grothendieck's period conjecture amounts here to the transcendence of π.

Other examples of periods include multiple zeta values

$$\sum_{n_1 > \cdots > n_k \geq 1} n_1^{-s_1} \ldots n_k^{-s_k} = \int_{1 \geq t_1 \geq \cdots \geq t_s \geq 0} \frac{dt_1}{\epsilon_1 - t_1} \cdots \frac{dt_s}{\epsilon_s - t_s}$$

($s_i \in \mathbb{Z}_{\geq 1}$, $; s = \sum s_i$, $\epsilon_j = 0$ or 1). Since Euler (who first considered such numbers), one has discovered a skein of algebraic relations between these numbers, and checked that all are of motivic origin. The underlying motivic Galois theory is now well understood thanks to the works of A. Goncharov and F. Brown[10] notably. Actually, it is the motivic approach which provides the best upper bound d_s (the best known, and conjecturally the best one) for the dimension of the $\mathbb{Q}$-space spanned by the multiple zeta values with fixed s: a Fibonacci-like recursion $d_s = d_{s-2} + d_{s-3}$.

---

[9] For all this, see *e.g.* the author's "Introduction aux Motifs" Panoramas et Synthèses 17, SMF 2004.

[10] "Mixed Tate Motives over Spec(Z)", in Annals of Math., volume 175, no. 1 (2012).

2.3. Fortunately, M. Kontsevich found an elementary and striking formulation of the period conjecture which does not involve motives[11].

Two fundamental rules of integral calculus are the Stokes formula and the formula of change of variable

$$\int_\Delta d\omega = \int_{\partial\Delta} \omega, \qquad \int_\Delta f^*\omega = \int_{f_*\Delta} \omega,$$

which provide immediately relations between (exponential) periods. The conjecture predicts that any algebraic relation (with algebraic coefficients) between (exponential) periods comes from these two rules.

For instance, Euler's identity $\zeta(2) = \pi^2/6$ can be understood as the identity of periods $\int_0^1 \int_0^1 \frac{2dxdy}{(1-xy)\sqrt{xy}} = (\int_0^\infty \frac{2dt}{1+t^2})^2$. E. Calabi has found how to derive it from a suitable change of variable: setting $x = u^2\frac{1+v^2}{1+u^2}$, $y = v^2\frac{1+u^2}{1+v^2}$, with Jacobian $|\frac{d(x,y)}{d(u,v)}| = \frac{4uv(1-u^2v^2)}{(1+u^2)(1+v^2)} = \frac{4(1-xy)\sqrt{xy}}{(1+u^2)(1+v^2)}$, one gets $\int_0^1 \int_0^1 \frac{2dxdy}{(1-xy)\sqrt{xy}} = \int\int_{u,v\geq 0,uv\leq 1} \frac{8dudv}{(1+u^2)(1+v^2)}$, which is equal to $(\int_0^\infty \frac{2dt}{1+t^2})^2$ (after the talk, C. Viola pointed out a more elementary way of deriving this identity, identifying $\zeta(2)$ to $\int_0^1 \int_0^1 \frac{dxdy}{(1-xy)}$, and using the change of variable $x = u - v, y = u + v$).

2.4. According to J. Ayoub[12], one could even dispense with the formula of change of variable: it follows from Stokes formula!

Let $\mathcal{A}$ be the algebra of functions of an arbitrary number of complex variables z_i, which are holomorphic on the polydisc $z_i \leq 1$, and algebraic over $\mathbb{Q}(z_1, \ldots, z_i, \ldots)$. One has a linear form $\mathcal{A} \xrightarrow{\int_\square} \mathbb{C}$ given by integration on the real hypercube $z_i \in [0, 1]$, and the point is to describe its kernel.

For any index i and any $g_i \in \mathcal{A}$, the function $h_i = \frac{\partial g_i}{\partial z_i} - g_{i|z_i=1} + g_{i|z_i=0}$ belongs to the kernel of $\int_\square$. Ayoub's reformulation of the period conjecture predicts that the kernel of $\int_\square$ is generated by such functions h_i.

For instance, starting from $h(z_1) \in \mathcal{A}$ and a algebraic function $f(z_1)$ which maps the unit disc (*resp.* the interval $[0, 1]$) into itself and fixes 0 and 1, the formula of change of variable shows that $f'(z_1)h(f(z_1)) - h(z_1)$ belongs to the kernel of $\int_\square$.

---

[11] *Cf.* M. Kontsevich, D. Zagier: Periods. Mathematics unlimited—2001 and beyond, 771–808, Springer.

[12] "Une version relative de la conjecture des périodes de Kontsevich-Zagier", to appear in Annals of Maths., http://www.math.polytechnique.fr/xups/vol11.html.

As Ayoub observed, one can write $f'(z_1)h(z_1) - h'(z_1)$ in the form predicted by the conjecture by setting $f_1 = f(z_1) - z_1$, $f_2 = -z_2 f'(z_1) + z_2 - 1$, and $g_i = f_i \cdot h(z_2 f(z_1) + (1 - z_2)z_1)$ for $i = 1, 2$.

Summarizing, one is led to the following speculation in line with the end of Galois's last letter: *the arithmetic of (exponential) periods should be dictated be elementary rules of integral calculus (and if so, could be described in terms of Galois groups).*

Simple things we don't know

Nicholas M. Katz

Abstract. This is a quite faithful rendering of a Colloquio De Giorgi I had the honor to give at Scuola Normale Superiore on March 21, 2012. The idea was to explain some open problems in arithmetic algebraic geometry which are simple to state but which remain shrouded in mystery.

1. An interactive game: dimension zero

Suppose I give you an integer $N \geq 2$, and tell you that I am thinking of a monic integer polynomial $f(X) \in \mathbb{Z}[X]$ whose discriminant $\Delta(f)$ divides some power of N. I tell you further, for every prime number p not[1] dividing N, the number

$$n_p(f) := \#\{x \in \mathbb{F}_p | f(x) = 0 \text{ in } \mathbb{F}_p\}$$

of its solutions in the prime field $\mathbb{F}_p := \mathbb{Z}/p\mathbb{Z}$. You must then tell me the degree of the polynomial f.

In this "infinite" version, where I tell you the $n_p(f)$ for every good prime, your task is simple; the degree of f is simply the largest of the $n_p(f)$. Indeed, $n_p(f) = \deg(f)$ precisely when p is a prime which splits completely in the number field $K_f := \mathbb{Q}(\text{the roots of f})$. By Chebotarev, this set of primes is infinite, and has density $1/\#Gal(K_f/\mathbb{Q})$.

If you do not have infinite patience, you may hope that you can specify a constant X_N, depending only on N, such that it will be enough for me to tell you $n_p(f)$ only for the good primes which are $\leq X_N$. Alas, this cannot be done. Whatever constant X_N you choose, I will pick an integer $a \geq 2$ such that $N^a > X_N$, and take for my f the cyclotomic polynomial $\Phi_{N^a}(X)$, whose roots are the primitive N^a'th roots of unity. With this choice of f, we have $n_p(f) = 0$ for all good primes $p \leq N^a$. Indeed, with this choice of f, $n_p(f)$ vanishes for a good prime p **unless** $p \equiv 1$

---

[1] We will call such a prime a good prime (for this problem).

mod N^a, in which case $n_p(f) = \deg(f)(= \phi(N^a) = N^{a-1}\phi(N)$, ϕ being Euler's ϕ function). But the condition that p be congruent to 1 mod N^a certainly forces $p > N^a$. In particular, we have $n_p(f) = 0$ for all good primes $p \leq N^a$.

2. An interactive game: curves

In this game, I give you an integer $N \geq 2$, and tell you that I am thinking of a (proper, smooth) curve $C/\mathbb{Z}[1/N]$ (with geometrically connected fibres). Once again I tell you, for every good prime p, the number

$$n_p(C) := \#C(\mathbb{F}_p),$$

the number of its points with values in $\mathbb{F}_p$. You must then tell me the common genus g of the geometric fibres of C.

What is your strategy? You know that, by Weil [10], the number $n_p(C)$ is approximately $p + 1$. More precisely, if we write

$$n_p(C) = p + 1 - a_p(C),$$

so that the data of the integers $n_p(C)$ is equivalent to the data of the integers $a_p(C)$, then we have the Weil bound

$$|a_p| \leq 2g\sqrt{p}.$$

A natural guess is that you can recover the integer $2g$ as the limsup of the ratios $|a_p|/\sqrt{p}$ as p varies over all good primes. You might even hope to recover $2g$ as the limsup of the ratios $a_p/\sqrt{p}$. Or you might make the more modest guess that you can recognize $2g$ as being the largest even integer such that, on the one hand, we have $|a_p|/\sqrt{p} \leq 2g$ **but** for at least one good prime we have $|a_p|/\sqrt{p} > 2g - 2$. Or you may be more ambitious and require that there are infinitely many good primes p with $|a_p|/\sqrt{p} > 2g - 2$.

The sad truth is that, except in some very special cases, none of these guesses is known to be correct. Let us first discuss the two cases where something is known, namely $g = 0$ and $g = 1$.

In the case of genus 0, then $a_p(C) = 0$ for every good p, and all guesses are hence correct.

In the case of genus one, the modest guess that we will have $a_p(C) \neq 0$ for infinitely many good p is easy to establish. First, we may replace our genus one curve C, which may not have a $\mathbb{Q}$-rational point, by its Jacobian, without changing the number of mod p points. Now $C/\mathbb{Z}[1/N]$ has a group(scheme) structure. In particular, each set $C(\mathbb{F}_p)$ has the structure of a finite abelian group. For any prime p not dividing $3N$ which splits completely in the number field $\mathbb{Q}(C[3] := $ the points of order 3), we know both that $C(\mathbb{F}_p)$ contains a subgroup of order 9, namely all the nine points of order dividing 3, and that p must be congruent to 1 mod 3 (this last fact because by the e_n pairing, once we have all the points of order any given n invertible in our field, that same field contains all the n'th roots of unity). So from the equality $n_p(C) = p + 1 - a_p(C)$, we get the congruences

$$a_p(C) \equiv p + 1 \mod 9, \quad p \equiv 1 \mod 3,$$

which together give the congruence

$$a_p(C) \equiv 2 \mod 3$$

for every prime p not dividing $3N$ which splits completely in $\mathbb{Q}(C[3])$.

In the $g = 1$ case, the truth of the Sato-Tate conjecture, established for non-CM[2] elliptic curves by Harris, Taylor et al., cf. [1], [2], [6], [9], leads easily to a proof that the most precise guess is correct in genus one. We will explain how this works in the next sections.

3. A "baby" version of the Sato-Tate conjecture

Let us begin with quick excursion into the world of compact Lie groups. For each even integer $2g \geq 2$, we denote by $USp(2g)$ the "compact symplectic group". We can see it concretely as the intersection of the complex symplectic group $Sp(2g, \mathbb{C})$ (take the standard symplectic basis $e_i, f_i, 1 \leq i \leq g$ in which $(e_i, e_j) = (f_i, f_j) = 0$ for all i, j and $(e_i, f_j) = \delta_{i,j}$) with the unitary group $U(2g)$ (where the same $e_i, f_i, 1 \leq i \leq g$ form an orthonormal basis). Or we can see $USp(2g)$ as a maximal compact subgroup of $Sp(2g, \mathbb{C})$, or we can see it as the "compact form" of $Sp(2g, \mathbb{C})$.

What is relevant here is that $USp(2g)$ is given to us with a $2g$-dimensional $\mathbb{C}$-representation, and in this representation every element has eigenvalues consisting of g pairs of complex conjugate numbers of absolute value one. Consequently, every element has its trace a real number which lies in the closed interval $[-2g, 2g]$.

For any closed subgroup K of $USp(2g)$, we also have a given $2g$-dimensional representation, whose traces lie in the closed interval $[-2g, 2g]$. Out of this data, we construct a "Sato-Tate measure" μ_K,

---

[2] In the CM case, Deuring proved that we are dealing with a Hecke character, and the required equidistribution for these goes back to Hecke.

a Borel probability measure on the closed interval $[-2g, 2g]$. Here are three equivalent descriptions of the measure μ_K. In all of them, we begin with the Haar measure $\mu_{\text{Haar},K}$ on K of total mass one. We have the trace, which we view as a continuous map

$$\text{Trace} : K \to [-2g, 2g].$$

In the fancy version, we define $\mu_K := \text{Trace}_*(\mu_{\text{Haar},K})$. More concretely, for any continuous $\mathbb{R}$-valued function f on the closed interval $[-2g, 2g]$, we impose the integration formula

$$\int_{[-2g,2g]} f d\mu_K := \int_K f(\text{Trace}(k)) d\mu_{\text{Haar},K}.$$

For an interval $I \subset [-2g, 2g]$, indeed for any Borel-measurable set $I \subset [-2g, 2g]$, its measure is given by

$$\mu_K(I) := \mu_{\text{Haar},K}(\{k \in K \mid \text{Trace}(k) \in I\}).$$

With these definitions in hand, we can state the "baby"[3] Sato-Tate conjecture.

Conjecture 3.1. Given an integer $N \geq 2$, and a projective smooth curve $C/\mathbb{Z}[1/N]$ with geometrically connected fibres of genus $g \geq 1$, there exists a compact subgroup $K \subset USp(2g)$ such that the sequence $\{a_p(C)/\sqrt{p}\}_{\text{good } p}$ is equidistributed in $[-2g, 2g]$ for the measure μ_K.

This equidistribution means that for any continuous $\mathbb{R}$-valued function f on the closed interval $[-2g, 2g]$, we have the integration formula

$$\int_{[-2g,2g]} f d\mu_K = \lim_{X \to \infty} (1/\pi_{\text{good}}(X)) \sum_{p \leq X, p \text{ good}} f(a_p(C)/\sqrt{p}),$$

where we have written $\pi_{\text{good}}(X)$ for the number of good primes up to X.

We now explain how the truth of this baby Sato-Tate conjecture for a given curve $C/\mathbb{Z}[1/N]$ implies that

$$2g = \text{limsum}_{\text{good } p} \ a_p(C)/\sqrt{p}.$$

We must show that for any real $\epsilon > 0$, there are infinitely many good primes p for which $a_p(C)/\sqrt{p}$ lies in the interval $(2g - \epsilon, 2g]$. For this

---

[3] "Baby" because it is the trace consequence of the "true" Sato-Tate conjecture, which we will not go into here. See [5] for a plethora of numerical evidence in the case $g = 2$.

we argue as follows. In any probability space, there are at most countably many points ("atoms") which have positive measure, cf. [4, page 135]. So at the expense of replacing the chosen ϵ by a smaller one, we may further assume that the point $2g - \epsilon$ is not an atom for the measure μ_K[4]. The the open set $(2g - \epsilon, 2g]$ has a boundary of measure zero, and hence [8, Proposition 1, I-19] we may apply the integration formula above to the characteristic function of this open set. Thus we get

$$\mu_K((2g - \epsilon, 2g]) = \lim_{X \to \infty} \frac{\#\{p \le X, p \text{ good}, a_p(C)/\sqrt{p} > 2g - \epsilon\}}{\#\{p \le X, p \text{ good}\}}.$$

But the set $\{k \in K, \text{Trace}(k) > 2g - \epsilon\}$ is open in K and contains the identity, so has strictly positive mass for Haar measure; this mass is, by definition, the μ_K measure of $(2g - \epsilon, 2g]$. Thus we have

$$\lim_{X \to \infty} \frac{\#\{p \le X, p \text{ good}, a_p(C)/\sqrt{p} > 2g - \epsilon\}}{\#\{p \le X, p \text{ good}\}} > 0,$$

and hence there are infinitely many good p for which $a_p(C)/\sqrt{p} > 2g - \epsilon$[5].

4. Integrality consequences

For any compact subgroup $K \subset U Sp(2g)$, the moments $M_{n,K} := \int_{[-2g,2g]} x^n d\mu_K$, $n \ge 0$, of the measure μ_K are nonnegative integers. Indeed, the n'th moment $M_{n,K}$ is the integral $\int_K (\text{Trace}(k))^n d\mu_{\text{Haar},K}$, which is the multiplicity of the trivial representation $\mathbb{1}$ in the n'th tensor power $std_{2g}^{\otimes n}$ of the given $2g$-dimensional representation std_{2g} of K.

---

[4] The point $2g$ is never an atom for the measure μ_K. Indeed, the only element of a compact subgroup $K \subset USp(2g)$ with trace $2g$ is the identity, and this point has positive measure in K if and only if K is finite. But if K were finite, then equidistribution would imply that $a_p(C)/\sqrt{p} = 2g$ for infinitely many good primes p. But the equality $a_p(C) = 2g\sqrt{p}$ holds for no p, simply because a_p is an integer, while $2g\sqrt{p}$ is not.

[5] If we keep our original ϵ, and allow the possibility that $2g - \epsilon$ is an atom, we can argue as follows. We take a continuous function f with values in $[0, 1]$ which is 1 on the interval $[2g - \epsilon/2, 2g]$ and which is 0 in $[-2g, 2g - \epsilon]$. Then we have the inequality

$$\liminf_{X \to \infty} \frac{\#\{p \le X, p \text{ good}, a_p(C)/\sqrt{p} > 2g - \epsilon\}}{\#\{p \le X, p \text{ good}\}} \ge$$

$$\ge \int_{[-2g,2g]} f d\mu_K \ge \int_{(2g-\epsilon/2, 2g]} f d\mu_K = \mu_K((2g - \epsilon/2, 2g]) > 0$$

and we conclude as above.

So the baby Sato-Tate conjecture predicts that for our curve $C/\mathbb{Z}[1/N]$, for each integer $n \geq 0$, the sums

$$(1/\pi_{\text{good}}(X)) \sum_{p \leq X, p \text{ good}} (a_p(C)/\sqrt{p})^n$$

not only have a limit as $X \to \infty$, but also that this limit is a nonnegative integer. Indeed, the baby Sato-Tate conjecture for $C/\mathbb{Z}[1/N]$ holds with a specified compact subgroup $K \subset USp(2g)$ if an only if[6] the above limit exist for each $n \geq 0$ and is equal to the n'th moment $M_{n,K}$.

5. Some test cases

There is a conjectural recipe for the compact subgroup $K \subset USp(2g)$ attached to a given $C/\mathbb{Z}[1/N]$ in terms of the ℓ-adic representation of $Gal(\overline{\mathbb{Q}}/\mathbb{Q})$ on the ℓ-adic Tate module of the Jacobian of $C/\mathbb{Z}[1/N]$. We will not go into that recipe here, except to say that it predicts that we will have $K = USp(2g)$ precisely when this ℓ-adic representation has an open image in the group $GSp(2g, \mathbb{Q}_\ell)$ of symplectic similitudes. A marvelous theorem [11] of Zarhin asserts that this is the case for any hyperelliptic curve $y^2 = h(x)$ with $h(x) \in \mathbb{Q}[x]$ a polynomial of degree $n \geq 5$ whose Galois group over $\mathbb{Q}$ is either the alternating group A_n or the full symmetric group S_n.

When $K = USp(2g)$, all the odd moments vanish, and one knows exact formulas for the first few even moments: $M_{0,USp(2g)} = 1$, and for $0 < 2k \leq 2g$ one has

$$M_{2k,USp(2g)} = 1 \times 3 \times \ldots \times (2k-1).$$

By a theorem of Schur, the truncated exponential series $e_n(x) := \sum_{0 \leq k \leq n} x^k/k!$ has Galois group A_n if 4 divides n, and S_n otherwise, cf. [3] for a beautiful exposition of Schur's theorem. Moreover, Coleman shows that the discriminant of $e_n(x)$ is $(-1)^{n(n-1)/2}(n!)^n$, so the only bad primes for $y^2 = e_n(x)$, whose genus is Floor$((n-1)/2)$, are those $p \leq n$. By a theorem of Osada [7, Corollary 3], the galois group of $h_n(x) := x^n - x - 1$ is S_n. The discriminant of $h_n(x)$ has much less regular behavior. It grows very rapidly, and often is divisible by huge primes[7]. So the bad primes for $y^2 = h_n(x)$, whose genus is Floor$((n-1)/2)$, are somewhat erratic.

---

[6] On a closed interval, a Borel probability measure is determined by its moments.

[7] For example, with $n = 17$, the discriminant is the prime 808793517812627212561, for $n = 22$ the prime factorization of the discriminant is $5 \times 6945409287652110798360556960$1.

In any case, for either of these curves, the baby Sato-Tate conjecture predicts that the sums

$$(1/\pi_{\mathrm{good}}(X)) \sum_{p \leq X, p \text{ good}} (a_p(C)/\sqrt{p})^d$$

tend to 0 for d odd, and for $d = 2k \leq 2g$ tend to

$$M_{2k, USp(2g)} = 1 \times 3 \times \ldots \times (2k - 1).$$

There is no integer $n \geq 5$ for which either of these statements is known, either for Schur's curve $y^2 = e_n(x)$ or for Osada's curve $y^2 = h_n(x)$.

Another striking but unknown consequence of baby Sato-Tate for these curves is this. Because the group $USp(2g)$ contains the scalar -1, the measure $\mu_{USp(2g)}$ on $[-2g, 2g]$ is invariant under $x \mapsto -x$. So for these two curves, the sets {good $p, a_p(C) > 0$} and {good $p, a_p(C) < 0$} should[8] each have Dirichlet density $1/2$.

Much remains to be done.

References

[1] T. BARNET-LAMB, D. GERAGHTY, M. HARRIS and R. TAYLOR, *A family of Calabi-Yau varieties and potential automorphy II*, Publ. Res. Inst. Math. Sci. **47** (2011), 29–98.

[2] L. CLOZEL, M. HARRIS and R. TAYLOR, *Automorphy for some ℓ-adic Lifts of Automorphic mod ℓ Galois Representations, with Appendix A, summarizing unpublished work of Russ Mann, and Appendix B by Marie-France Vignéras*, Publ. Math. Inst. Hautes Études Sci. **108** (2008), 1–181.

[3] R. COLEMAN, *On the Galois groups of the exponential Taylor polynomials*, Enseign. Math. **33** (1987), 183–189.

[4] W. FELLER, "An Introduction to Probability Theory and its Applications", Vol. II, John Wiley and Sons, 1966.

[5] F. FITÉ, K. KEDLAYA, V. ROTGER and A. SUTHERLAND, *Sato-Tate distributions and Galois endomorphism modules in genus 2*, Compos. Math. **148** (2012), 1390–1442.

[6] M. HARRIS, N. SHEPHERD-BARRON and R. TAYLOR, *A family of Calabi-Yau varieties and potential automorphy*, Ann. of Math. **171** (2010), 779–813.

[7] H. OSADA, *The Galois groups of the polynomials $X^n + aX^\ell + b$*, J. Number Th. **25** (1987), 230–238.

---

[8] The measure $\mu_{USp(2g)}$ has no atoms. In particular 0 is not an atom.

[8] J.-P. SERRE, "Abelian ℓ-adic Representations", Addison-Wesley, 1989.

[9] R. TAYLOR, *Automorphy for some ℓ-adic lifts of automorphic modℓ Galois representations*, II, Publ. Math. Inst. Hautes Études Sci. **108** (2008), 183–239.

[10] ANDÉ WEIL, "Variétés abéliennes et courbes algébriques", Actualités Sci. Ind., no. 1064, Publ. Inst. Math. Univ. Strasbourg **8** (1946), Hermann & Cie., Paris, 1948. 165 pp.

[11] YURI G. ZARHIN, *Very simple 2-adic representations and hyperelliptic Jacobians*, Mosc. Math. J. **2** (2002), 403–431.

Is laziness paying off?
("Absorbing" method)

Endre Szemerédi

Abstract. We are going to mention some embedding problems. The main ideas of most of the solution of these problems are that do not work too much. Finish the job almost, then use some kind of absorbing configuration (that is what we call laziness).

1. Introduction

We are going to discuss problems both for graphs and hypergraphs.

2. Graphs

Let $\mathcal{G} = (V(\mathcal{G}), E(\mathcal{G}))$ be a graph, $|V(\mathcal{G})| = n$ (For a set A $|A|$ denotes the number of elements of A.) In our discussion every set will be finite. For $x \in V(\mathcal{G})$ $N(x) = \{y : (x, y) \in E(\mathcal{G})\}$, $d(x) = |N(x)|$, $\delta(\mathcal{G}) = \min d(x)$ $x \in V(\mathcal{G})$. It is well known that if $\delta(\mathcal{G}) \geq \frac{n}{2}$, then $\mathcal{G}$ contains a one-factor ($|V(\mathcal{G})| = n$), where a 1-factor is a set of $n/2$ independent edges, (see Figure 1).

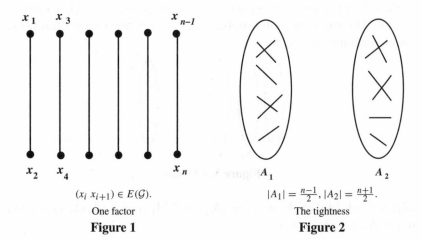

$(x_i \, x_{i+1}) \in E(\mathcal{G}).$

One factor

Figure 1

$|A_1| = \frac{n-1}{2}, |A_2| = \frac{n+1}{2}.$

The tightness

Figure 2

This result is tight. For the tightness see Figure 2: Our graph is the union of the complete graph with vertex set A_1 and the complete graph with vertex set A_2. Can we say something more interesting if $\delta(\mathcal{G}) \geq \frac{n}{2}$? The answer is yes. The classical result of Dirac [7] states that a sufficient condition for n-vertex graph to be Hamiltonian is that the minimum degree is at least $\frac{n}{2}$.

The previous counterexample (Figure 2) shows that Dirac's result is tight. (A graph $\mathcal{G} = \{V(\mathcal{G}), E(\mathcal{G})\}$ contains a Hamiltonian cycle if its vertices can be arranged such that between two consecutive vertices there is an edge of $\mathcal{G}$. The order of the vertices: $v_1, v_2, v_3, \ldots, v_n$.

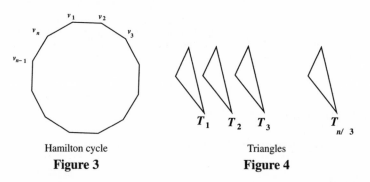

Hamilton cycle

Figure 3

Triangles

Figure 4

What can we say if we modify the minimum degree condition? If $\delta(\mathcal{G}) \geq \frac{2}{3}n$, then the theorem of Corradi and Hajnal [4] states that the graph contains vertex disjoint triangles so that the union of their vertex set is $V(\mathcal{G})$, see Figure 4. (Again we assume that $n = 3\ell$.)

Erdős conjectured (1962) that if $\delta(\mathcal{G}) \geq \frac{k-1}{k}$, then $\mathcal{G}$ contains vertex disjoint complete graph on k vertices such that the union of their vertex set is $V(\mathcal{G})$. (Again we assume that $n = k \cdot \ell$.)

Hajnal and Szemerédi [8] proved Erdős's conjecture. The result of Corradi, Hajnal, Szemerédi is tight, which can be shown by the following counterexample, on Figure 5.

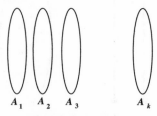

Figure 5. k blocks.

$|A_1| = \frac{n}{k} - 1, |A_2| = A_3 = \ldots = |A_{k_1}| = \frac{n}{k}, A_{k+1} = \frac{n}{k} + 1, (x, y) \in E(y)$ if $x \in A_i, y \in A_j, i \neq j$.

El-Zahar's Conjecture (1984)

$$\mathcal{G} = \{V(\mathcal{G}), E(\mathcal{G})\}, \quad |V(\mathcal{G})| = n$$

F is a family of graphs consisting of r vertex disjoint cycles of length $n_1, n_2, \ldots, n_r$ such that

$$\sum_{i=1}^{r} n_i = n \quad \text{and} \quad \delta(\mathcal{G}) \geq \sum_{i=1}^{r} \left\lceil \frac{n_i}{2} \right\rceil. \tag{1}$$

The conjecture states that inequality (1) implies that $\mathcal{G}$ contains F as a subgraph. El Zahar's conjecture was proved by S. Abbasi, I. Khan, G. Sárközy, E. Szemerédi [1]. Special cases of the conjecture for instance are

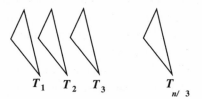

Figure 6. Triangles, Corradi–Hajnal.

Figure 7. 4-tuples, Erdős–Faudree configuration.

Erdős–Fadree's conjecture was proved by Cs. Magyar.

Pósa–Seymour Conjecture

We are going to discuss Pósa's conjecture and Seymour's conjecture. We remark that Pósa's conjecture is a special case of Seymour's conjecture. We have to define the square of a Hamilton cycle and the k-th power of a Hamilton cycle.

The square of a Hamilton cycle is when we have an ordering of the vertices, say $x_1, \ldots, x_n$ and we join x_i to x_{i+1} and x_{i+2}, where the indices are counted mod n (Figure 8). The k-th power of a Hamilton cycle is when we have an ordering of the vertices, as above: $x_1, \ldots, x_n$ and we join x_i to $x_{i+1}, x_{i+2}, \ldots, x_{i+k-1}$, where again, the indices are counted mod n (for $k = 3$ see Figure 9).

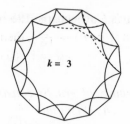

Square of a Hamiltonian cycle	Cube of the Hamiltonian cycle.
Figure 8	**Figure 9**

Pósa's conjecture states that if $\delta(\mathcal{G}) \geq \frac{2}{3}n$, then the graph $\mathcal{G}$ contains a square of a Hamilton cycle. Seymour's conjecture states that if $\delta(\mathcal{G}) \geq \frac{k-1}{k}n$, then the graph $\mathcal{G}$ contains a k-th power of a Hamilton cycle.

J. Komlós, G. Sárközy, E. Szemerédi [16, 17] proved both conjecture, but their proof used the regularity and the blow-up lemmas, therefore n is very large in their proof. The new proof of A. Jasmed, E. Szemerédi [11] for the Seymour's conjecture did not use the regularity lemma. They introduced a new type of "connecting" lemma and an absorbing lemma.

Is Pósa's conjecture true for every n?

Presently $n \leq 300\,000\,000$. This is the number of population in U.S.A. This was proved by Chau, DeBiasio, and Kierstead [3] in 2011. With some work it can be reduced to the number of population of a much smaller country (work in progress).

Packing trees into a graph $\mathcal{G}$

$\Delta(T)$ denotes the maximum degree of a tree T on n vertices. Bollobás conjectured that if $\delta(\mathcal{G}) \geq (1 + \varepsilon)\frac{n}{2}$ and $\Delta(T) < c$, then if n is large enough, $\mathcal{G}$ contains T.

Bollobás's conjecture was proved by J. Komlós, G. Sárközy, E. Szemerédi [17, 18].

One can ask whether we need that $\delta(\mathcal{G}) \geq \left(\frac{1}{2} + \varepsilon\right)n$. The answer is no, which was proved by B. Csaba, I. Levitt, J. Nagy-György, E. Szemerédi [5].

They proved that if $\Delta(T) < C$ and $\delta(\mathcal{G}) > \frac{n}{2} + f(C)\log n$, then $\mathcal{G}$ contains the tree T. Their result is tight, which is shown by the following example:

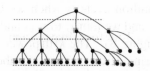

Figure 10. Our graph is a complete binary tree.

Again we can ask whether we need $\Delta(T) < C$.

No, but if $\Delta(T) > \frac{2n}{\log n}$, then $T \not\subset \mathcal{G}$ which is shown by the following example

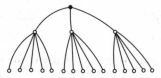

Figure 11. Degrees: $d(x_1) = \frac{1}{2} \log n$, and $d(x_i) = \frac{2n}{\log n}$.

Our graph $\mathcal{G}$ is a random graph with minimal degree $(1 - \delta)n$. B. Csaba, A. Jashmed, E. Szemerédi [6] proved that if

$$\Delta(T) < \frac{1}{2} \frac{n}{\log n} \quad \text{and}$$

$$\delta(\mathcal{G}) > \frac{n}{2} + 2\Delta(T) \log n, \quad \text{then } T \subset \mathcal{G}.$$

As we saw at the two ends the theorem is tight. What is in between?

Remark 2.1. If $m = o(\sqrt{n})$ and $\mathcal{G}$ is not "extremal", then $\delta(\mathcal{G}) \geqq \left(\frac{1}{2} - \varepsilon\right) n$ is enough.

The extremal graph is the following

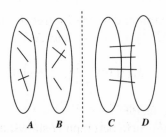

Figure 12. A and B are almost complete graphs $K_{n/2}$. C and D form an almost complete bipartite graph $|A| = |B| = |C| = |D| = n/2$.

3. Hypergraphs

$\mathcal{H} = \{V(H), E(H)\}$, is a 3-uniform hypergraph. For $x \in V(\mathcal{H})$

$$N^*(x) = \{(u, w); \ (x, v, w) \in \mathcal{H}\}$$

$$\delta_2^*(\mathcal{H}) = \min_{x \in V(\mathcal{H})} |N^*(x)|.$$

We are interested in a good bound for $\delta_2^*(\mathcal{H})$ such that $\mathcal{H}$ contains a one-factor (Hamiltonian cycle). The the following hypergraph (Figure 13) shows that $\delta_2^*(\mathcal{H}) \geqq \frac{5}{9}\frac{n^2}{2}$ $\mathcal{H}$:

$$|A| = \tfrac{1}{3}n - 1$$
$$|B| = \tfrac{2}{3}n + 1$$

Obviously $\delta_2^*(\mathcal{H}) = \frac{5}{9}\frac{n^2}{2}$. It is easy to see that $\mathcal{H}$ does not contain a one-factor. $\{x, y, z\} \in \mathcal{H}$ if at least one of x, y, z is in A.

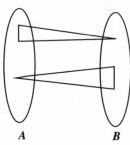

A   B

Figure 13.

H. Han, Y. Person, M. Schacht [10] proved that there is a one-factor if

$$\delta_2^*(\mathcal{H}) > \left(\frac{5}{9} + \varepsilon\right)\binom{n}{2}.$$

D. Kühn, Osthus and Treglown [22] and I. Khan [14] proved that $\mathcal{H}$ contains a one-factor if $\delta_2^*(\mathcal{H}) \geq \frac{5}{9}\binom{n}{2}$.
$\mathcal{H}$ is a 4-uniform hypergraph for $x \in V(\mathcal{H}) N^*(x) = \{(u,v,w); (x,u,v,w) \in \mathcal{H}\}$

$$\delta_3^*(\mathcal{H}) = \min_{x \in V(\mathcal{H})} |N^*(x)|.$$

I. Khan [15] proved that if

$$\delta_3^*(\mathcal{H}) \geqq \frac{37}{64}\binom{n-1}{3},$$

then $\mathcal{H}$ contains a one-factor.

The following 4-uniform hypergraph $\mathcal{H}$ shows that the theorem is tight.

$$|A| = \tfrac{1}{4}n - 1$$
$$|B| = \tfrac{3}{4}n + 1$$

$(x, y, z, w) \in \mathcal{H}$ if at least one vertex is in A. It is easy to see that $\mathcal{H}$ does not contain a one-factor.

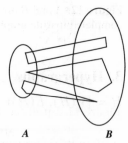

A   B

Figure 14.

N. Alon, P. Frankl, H. Huang, V. Rödl, A. Ruciński, B. Sudakov [2] proved that for 5-uniform hypergraph if $\delta_4^*(\mathcal{H}) \geqq \left(1 - \left(\frac{4}{5}\right)^4\right)(1+\varepsilon)\binom{n-1}{4}$ then $\mathcal{H}$ contains a one-factor.

We are going to formulate a general conjecture.

Let $\mathcal{H}$ be a k-uniform hypergraph

$$N^*(x) = \{(x_1, x_2, \ldots, x_{k-1}); (x_1, x_2, \ldots, x_{k-1}x) \in \mathcal{H}\}$$
$$\delta_{k-1}^*(\mathcal{H}) = \min_{x \in V(\mathcal{H})} |N^*(x)|.$$

The general conjecture is the following:

If $\delta_{k-1}^*(\mathcal{H}) \geq \left(1 - \left(\frac{k-1}{k}\right)^{k-1}\right)\binom{n-1}{k-1}$ $(|V(\mathcal{H})| = n)$, then $\mathcal{H}$ contains a one-factor.

By my opinion it is one of the most important conjecture in extremal hypergraph theory.

Hamiltonian cycles After discussing the one-factor problems we are going to discuss different Hamiltonian cycle problems.

Let $\mathcal{H}$ be a k-uniform hypergraph. For $x_1, x_2, \ldots, x_{k-1} \in V(\mathcal{H})$

$$N^*(x_1, x_2, \ldots, x_{k-1}) = \{y; (x_1, x_2, \ldots, x_{k-1}, y) \in \mathcal{H}\}$$
$$\delta_{k-1}^*(\mathcal{H}) = \min_{x_1, x_2, \ldots, x_{k-1} \in V(\mathcal{H})} N^*(x_1, x_2, \ldots, x_{k-1}).$$

For a k-uniform hypergraph $\mathcal{H}$ by a $k - 1$-cycle of length $l \geq k + 1$ we mean a k-uniform hypergraph whose vertices can be ordered cyclically $x_1, x_2, \ldots, x_l$ in such a way that for each $i = 1, 2, \ldots, l$ the set $\{v_i, v_{i+1}, \ldots, v_{i+k-1}\}$ is an edge, where for $h > l$ we set $x_h = x_{h-l}$

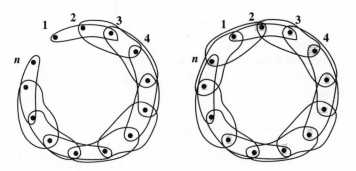

Figure 15. 2-path and 2-cycle of length 13.

A Hamiltonian cycle in a k-uniform hypergraph $\mathcal{H}$ is a sub-hypergraph of $\mathcal{H}$ which is a $(k - 1)$-cycle and contains all vertices of $\mathcal{H}$.

H. Kierstead and Gy. Katona [12] proved that if

$$\delta_{k-1}^*(\mathcal{H}) \geq \left(\frac{1}{2} - \frac{1}{2k}\right) \eta$$

then $\mathcal{H}$ contains a Hamiltonian cycle.

They conjectured that if $\delta_{k-1}^*(\mathcal{H}) \geq \frac{1}{2}\eta$, then $\mathcal{H}$ contains a Hamiltonian cycle.

If the conjecture is true, then it is tight, which is shown by the following construction:

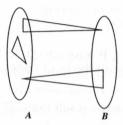

Figure 16. $|A| = |B| = \frac{n}{2}$ and $\{x, y, z\} \in \mathcal{H}$ if at least one of x, y, z is in A.

It is a stable construction, meaning that if a hypergraph $\mathcal{H}$ has a different structure than the hypergraph in the above construction, then

$$\delta_{k-1}^*(\mathcal{H}) \geq \left(\frac{1}{2} - \varepsilon\right) n$$

is enough.

V. Rödl, A. Ruciński, E. Szemerédi [24] proved that if

$$\delta_2(\mathcal{H}) > \left(\frac{1}{2} + \varepsilon\right) n,$$

then $\mathcal{H}$ contains a Hamiltonian cycle.

Later in [25] they proved that if $\delta_{k-1}^*(\mathcal{H}) \geq \left(\frac{1}{2} + \varepsilon\right) n$, then $\mathcal{H}$ contains a Hamiltonian cycle.

In [26] they proved the exact result. They proved that if $\delta_2(\mathcal{H}) \geq \frac{1}{2}n$, then $\mathcal{H}$ contains a Hamiltonian cycle. Moreover, there is only one example where the result is tight. This is in Figure 16.

The proof is much more difficult than the approximate version.

$$\left(\delta_2^*(\mathcal{H}) \geq \left(\frac{1}{2} + \varepsilon\right) n.\right)$$

(Work in progress for $\delta_{k-1}^*(\mathcal{H})$.)

For a k-uniform hypergraph $\mathcal{H}$ and an integer m by an m cycle of length l we mean a k-uniform hypergraph whose vertices can be ordered cyclically $x_1, x_2, \ldots, x_l$ in such a way that $(x_1, x_2, \ldots, x_k) \in \mathcal{H}$,

$$(x_{k-m-1}, x_{k-m+1}, x_k, x_{k+1}, \ldots, x_{2k-m}) \in \mathcal{H}$$

and if $(x_j, x_{j+1}, \ldots, x_{j+k-1}) \in \mathcal{H}$, then $(x_{j+k-1-m-1}, \ldots, x_{j+2k-1-m}) \in \mathcal{H}$.

Figure 17. $m = 3$, and $k = 8$.

D. Kühn, D. Osthus [21], P. Keevash, D. Kuhn, R. Mycroft, D. Osthus [13], H. Han, M. Schacht [9], D. Kühn, R. Mycroft, D. Osthus [20] have got several results about m Hamiltonian cycle in k-uniform hypergraphs.

Let me mention here only a result from [20].

If $\delta^*_{k-1}(\mathcal{H}) = \left(\frac{1}{\lceil \frac{k}{k-m} \rceil (k-m)} + o(1) \right) n$, then $\mathcal{H}$ contains an m Hamiltonian cycle.

Lo and Markström [23] used the absorption method for many problems. Let me mention just two of them.

1-factors in 3-partite 3-uniform graphs. (The multipartite version of the Hajnal–Szemerédi theorem).

F-factors in hypergraphs.

4. Absorbing Lemmas

Now we are going to mention some absorbing configurations and some absorbing lemmas.

For the El-Zahar's problem the absorbing configuration is the following:

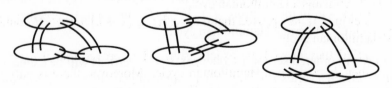

Figure 18. $\gamma \frac{n}{\log n}$ copies of $K_3(t)$, where $t = \log n$.

For the tree-packing problem the absorbing configuration is simple. We randomly chose three subsets, M, D_1, and D_2 of the vertex set $V(\mathcal{G})$ (see Figure 19).

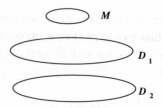

Figure 19. The sizes of the three groups are: $|M| = \gamma^3 n$, $|D_1| = \gamma n$, and $|D_2| = \gamma n$.

Discussing two absorbing lemmas in details. To state and prove the absorbing lemma from [24] we need a definition and a lemma which we call a connecting lemma. For two paths P and Q let ab be an endpair of P and ba be an endpair of Q, and assume further that

$$V(P) \cap V(Q) = \{ab\}.$$

By $P \circ Q$ we denote the path obtained (in a unique way) as a concatenation of P and Q. This definition extends naturally to more than two paths.

Connecting lemma. *For every two disjoint and ordered pairs of vertices* (x, y) *and* (c, d) *there is a 3-path of length* $= m$ *(*$m < \frac{4}{\gamma}$, γ *is small) in* $\mathcal{H}$ *which connects* (x, y) *and* (c, d).

Absorbing lemma. *There is an l-path A in H with* $l = |V(A)| \leq 20\gamma^2 n$ *such that for every subset* $U \subset V \setminus V(A)$ *of size at most* $\gamma^5 n$ *there is a path* A_U *in H with* $V(A_U) = V(A) \cup U$ *and such that* A_U *has the same endpairs as A.*

In other words, this lemma asserts that there is *one* not too long path such that *every* not too large subset can be "absorbed" into this path by creating a longer path with the same endpoints.

Consequently, if this path happens to be a segment of a cycle C of order at least $(1 - \gamma^5)n$ then, setting $U = V \setminus V(C)$, the path A_U together with the path $C \setminus A$ form a Hamiltonian cycle.

Proof. An ordered set (or a sequence) of four vertices will be called a 4-*tuple*. Given a vertex v we say that a 4-tuple of vertices x, y, z, w *absorbs* v if $xyz, yzw, xyv, yvz, vzw \in H$. A 4-tuple is called *absorbing* if it absorbs a vertex. This terminology reflects the fact that the path $xyzw$ can be extended by inserting (or absorbing) vertex v to create the path $xyvzw$. Note that both paths have the same set of endpoints. $\square$

Claim 1. *For every* $v \in V$ *there are at least* $2\gamma^2 n^4$ *4-tuples absorbing* v.

Proof. Because H is an (n, γ)-graph, there are at least $(n-1)(1/2+\gamma)n$ ordered pairs yz such that $vyz \in H$. For each such pair there are at least $2\gamma n$ common neighbours x of vy and yz, and at least $2\gamma n - 1$ common neighbours w of vz and yz, yielding together at least

$$(n-1)(1/2+\gamma)n2\gamma n(2\gamma n - 1) > 2\gamma^2 n^4$$

4-tuples absorbing v. □

For each $v \in V$, let $\mathcal{A}_v$ be the family of all 4-tuples absorbing v. The next claim is obtained by the probabilistic method.

Claim 2. *There exists a family $\mathcal{F}$ of at most $2\gamma^3 n$ disjoint, absorbing 4-tuples of vertices of H such that for every $v \in V$, $|\mathcal{A}_v \cap \mathcal{F}| > \gamma^5 n$.*

Proof. We first select a family $\mathcal{F}'$ of 4-tuples at random by including each of $n(n-1)(n-2)(n-3) \sim n^4$ of them independently with probability $\gamma^3 n^{-3}$ (some of the selected 4-tuples may not be absorbing at all). By Chernoff's inequality, with probability $1 - o(1)$, as $n \to \infty$,

- $|\mathcal{F}'| < 2\gamma^3 n$, and
- for each $v \in V$, $|\mathcal{A}_v \cap \mathcal{F}'| > \frac{3}{2}\gamma^5 n$.

Moreover, the expected number of intersecting pairs of 4-tuples in $\mathcal{F}'$ is at most

$$n^4 \times 4 \times 4 \times n^3 \times (\gamma^3 n^{-3})^2 = 16\gamma^6 n,$$

and so, by Markov's inequality, with probability at least $1/17$

- there are at most $17\gamma^6 n$ pairs of intersecting 4-tuples in $\mathcal{F}'$.

Thus, with positive probability, a random family $\mathcal{F}'$ possesses all three properties marked by the bullets above, and hence there exists at least one such family which, with a little abuse of notation, we also denote by $\mathcal{F}'$.

After deleting from $\mathcal{F}'$ all 4-tuples intersecting other 4-tuples in $\mathcal{F}'$, as well as those which do not absorb any vertex, we obtain a subfamily $\mathcal{F}$ of $\mathcal{F}'$ consisting of disjoint and absorbing 4-tuples and such that for each $v \in V$,

$$|\mathcal{A}_v \cap \mathcal{F}| > \frac{3}{2}\gamma^5 n - 34\gamma^6 n > \gamma^5 n.$$ □

Set $f = |\mathcal{F}|$ and let $F_1, \ldots, F_f$ be the elements of $\mathcal{F}$. For each $i = 1, \ldots, f$, F_i is absorbing and thus spans a 4-path in H. We will further denote these paths also by F_i and set $F = \bigcup_{i=1}^{f} F_i$.

Our next task is to connect all these 4-paths into one, not too long path A. To this end, we will repeatedly apply the Connecting lemma and, for each $i = 1, \ldots, f - 1$, connect the endpairs of F_i and F_{i+1} by a short path.

Claim 3. *There exists a path A in H of the form*

$$A = F_1 \circ C_1 \circ \cdots \circ F_{f-1} \circ C_{f-1} \circ F_f$$

where the paths $C_1, \ldots, C_{f-1}$ have each at most $8/\gamma$ vertices.

Proof. We will prove by induction on i that for each $i = 1, \ldots, f$, there exists a path A_i in H of the form $A_1 = F_1$ and, for $i \geq 2$,

$$A_i = F_1 \circ C_1 \circ \cdots \circ F_{i-1} \circ C_{i-1} \circ F_i,$$

where the paths $C_1, \ldots, C_{i-1}$ have each at most $8/\gamma$ vertices.
Then $A = A_f$.
There is nothing to prove for $i = 1$. Assume the statement is true for some $1 \leq i \leq f - 1$. Let ab be an endpair of A_i and let cd be an endpair of F_{i+1}. Denote by H_i the subhypergraph induced in H by the set of vertices $V_i = (V \setminus V(F \cup A_i)) \cup \{a, b, c, d\}$. Since

$$|V(F \cup A_i)| < |\mathcal{F}|(4 + 8/\gamma) < 10f/\gamma < 20\gamma^2 n,$$

H_i is a $(|V_i|, \gamma/2)$-graph, where $0 < n - |V_i| < 20\gamma^2 n$. By the Connecting lemma applied to H_i and the pairs ba and dc, there is a path $C_i \subset H_i$ of length at most $4/(\gamma/2) = 8/\gamma$, connecting these pairs.
Note that $V(C_i) \setminus \{a, b, c, d\}$ is disjoint from $V(F \cup A_i)$, and thus,

$$A_{i+1} = A_i \circ C_i \circ F_{i+1}$$

is the desired path.   □

Claim 3 states that we may connect all 4-paths in $\mathcal{F}$ into one path A of length at most $f(4 + 8/\gamma) < 20\gamma^2 n$. It remains to show that A has the absorbing property. Let $U \subset V \setminus V(A)$, $|U| \leq \gamma^5 n$. Because for every $v \in U$ we have $|\mathcal{A}_v \cap \mathcal{F}| > \gamma^5 n$, that is, there are at least $\gamma^5 n$ disjoint, v-absorbing 4-tuples in A, we can insert all vertices of U into A one by one, each time using a fresh absorbing 4-tuple.   □

In [10] a very powerful absorbing lemma was formulated and proved. Before formulating this lemma, we prove an easy proposition.

Proposition 1. *Let $\mathcal{H}$ be a k-uniform hypergraph on n vertices. For all $x \in [0, 1]$ and all integers $m \leq \ell$ we have, if*

$$\delta_\ell(\mathcal{H}) \geq x \binom{n}{k-\ell}, \quad \text{then} \quad \delta_m(\mathcal{H}) \geq x \binom{n}{k-m} - O(n^{km1}),$$

where the constant in the error term only depends on k, ℓ, and m.

Proof. Consider an arbitrary m-set $T = \{v_1, \ldots, v_m\} \in \binom{V(\mathcal{H})}{m}$. Then the condition on $\delta_\ell(\mathcal{H})$ implies that T is contained in at least

$$\binom{k-m}{\ell-m}^{-1} \sum_{v_{m+1}, \ldots, v_\ell \in \binom{V \backslash T}{\ell-m}} \deg(v_1, \ldots, v_\ell) \geq \binom{k-m}{\ell-m}^{-1} \binom{n-m}{\ell-m} x \binom{n}{k-\ell}$$

$$\geq x \binom{n}{k-m} - O(n^{k-m-1})$$

edges, and the proposition follows. $\qquad\qquad\qquad\qquad\qquad\qquad\square$

Now we can formulate and prove our lemma.

Absorbing lemma. *For all $\gamma > 0$ and integers $k > \ell > 0$ there is an n_0 such that for all $n > n_0$ the following holds: Suppose $\mathcal{H}$ is a k-uniform hypergraph on n vertices with minimum ℓ-degree $\delta_\ell(\mathcal{H}) \geq (1/2 + 2\gamma)\binom{n}{k-\ell}$, then there exists a matching M in $\mathcal{H}$ of size $|M| \leq \gamma^k n/k$ such that for every set $W \subset V \setminus V(M)$ of size at most $\gamma^{2k} n \geq |W| \in k\mathbb{Z}$ there exists a matching covering exactly the vertices in $V(M) \cup W$.*

Proof. Let $\mathcal{H}$ be a k-uniform hypergraph with $\delta_\ell(\mathcal{H}) \geq (1/2 + 2\gamma)\binom{n}{k-\ell}$. From Proposition 1 we know $\delta_1(\mathcal{H}) \geq \left(\frac{1}{2} + \gamma\right)\binom{n}{k-1}$ (for all large n) and it suffices to prove the lemma for $\ell = 1$.

Throughout the proof we assume (without loss of generality) that $\gamma \leq 1/10$ and let n_0 be chosen sufficiently large. Further set $m = k(k-1)$ and call a set $A \in \binom{V}{m}$ of size m an **absorbing** m-set for $T = \{v_1, \ldots, v_k\} \in \binom{V}{k}$ if A spans a matching of size $k - 1$ and $A \cup T$ spans a matching of size k, i.e., $\mathcal{H}[A]$ and $\mathcal{H}[A \cup T]$ both contain a perfect matching. $\qquad\square$

Claim 4. *For every $T = \{v_1, \ldots, v_k\} \in \binom{V}{k}$ there are at least $\gamma^{k-1}\binom{n}{k-1}^k/2$ absorbing m-sets for T.*

Thus, for $X_{i,J} = X_{i,J}(T) = |\{E \in M_i : E \text{ survived}\}|$ we have

$$\mu_{i,J} = \mu_{i,J}(T) = \mathbb{E}[X_{i,J}] = \frac{m!}{k^m}|M_i|.$$

Now call a matching M_i bad (with respect to the chosen partition $U_1, ..., U_k$) if there exists a set $J \in \binom{[k]}{m}$ such that

$$X_{i,J} \leq \left(1 - \left(\frac{(4k2)\ln n}{\mu_{i,J}}\right)^{1/2}\right)\mu_{i,J}$$

and call T a bad set (with respect to $U_1, \ldots, U_k$) if there is at least one bad $M_i = M_i(T)$. Otherwise call T a good set. For a fixed M_i the events "E survived" with $E \in M_i$ are jointly independent, hence we can apply Chernoffs inequality and we obtain

$$\mathbb{P}[M_i \text{ is bad}] \leq \binom{k}{m}\exp((2k1)\ln n) = \binom{k}{m}n^{2k+1}.$$

Summing over all matchings M_i and recalling $i_0 \leq mn^{m1}$ and $m \leq k1$ yields

$$\mathbb{P}[\text{there is at least one bad } M_i] \leq i_0\binom{k}{m}n^{2k+1} \leq n^k$$

and summing over all ℓ-sets T we obtain

$$\mathbb{P}[\text{there is at least one bad } T] \leq n^\ell n^k \leq n^1.$$

Moreover, Chernoffs inequality yields

$$\mathbb{P}\left[\exists k_0 \in [k] : |U_{k_0}| > n/k + n^{1/2}(\ln n)^{1/4}/k\right] \leq k\exp(0(\ln n)^{1/2}/(3k))$$
$$= o(1).$$

Thus, with positive probability there is a partition $U_1, \ldots, U_k$ such that all ℓ-sets T are good and such that

$$|U_j| \leq n/k + n^{1/2}(\ln n)^{1/4}/k \text{ for every } j \in [k].$$

Consequently, by redistributing at most $n^{1/2}(\ln n)^{1/4}$ vertices of the partition $U_1, \ldots, U_k$ we obtain an equipartition partition $V = V_1\dot\cup\ldots\dot\cup V_k$ with

$$|V_j| = n/k \text{ and } |U_j \setminus V_j| \leq n^{1/2}(\ln n)^{1/4}/k \text{ for every } j \in [k].$$

To verify that the partition $V_1, \ldots, V_k$ satisfies the claim of the lemma note that for a crossing ℓ set T and the m-set $J = \{j \in [k] : T \cap V_j = \emptyset\}$

we have

$$\deg(T) \geq \sum_{i \in [i_0]} \left(1 - \left(\frac{(4k-2)\ln n}{\mu_{i,J}(T)}\right)^{1/2}\right)\mu_{i,J}(T) - m\frac{n^{1/2}(\ln n)^{1/4}}{k}n^{m-1}$$

$$\geq \sum_{i \in [i_0]} \mu_{i,J}(T) - ((4k-2)\ln n)^{1/2}\sum_{i \in [i_0]}(\mu_{i,J}(T))^{1/2}$$

$$-(\ln n)^{1/4}n^{m-1/2}$$

$$= \frac{m!}{k^m}\deg(T) - ((4k-2)\ln n)^{1/2}\sum_{i \in [i_0]}(\mu_{i,J}(T))^{1/2}$$

$$-(\ln n)^{1/4}n^{m-1/2}.$$

The Cauchy–Schwarz inequality then gives

$$\sum_{i \in [i_0]}(\mu_{i,J}(T))^{1/2} \leq \left(i_0 \sum_{i \in [i_0]}\mu_{i,J}(T)\right)^{1/2} \leq \left(mn^{m-1}\binom{n}{m}\right)^{1/2} \leq n^{m-1/2}.$$

Proof. Let $T = \{v_1, \ldots, v_k\}$ be fixed. Since n_0 was chosen large enough there are at most $(k-1)\binom{n}{k-2} \leq \gamma\binom{n}{k-1}$ edges which contain v_1 and v_j for some $j \in \{2, \ldots, k\}$. Due to the minimum degree of $\mathcal{H}$ there are at least $\binom{n}{k-1}/2$ edges containing v_1 but none of the vertices $v_2, \ldots, v_k$. We fix one such edge $\{v_1, u_2, \ldots, u_k\}$ and set $U_1 = \{u_2, \ldots, u_k\}$. For each $i = 2, 3, \ldots, k$ and each pair u_i, v_i suppose we succeed to choose a set U_i such that U_i is disjoint to $W_{i-1} = \bigcup_{j \in [i-1]} U_j \cup T$ and both $U_i \cup \{u_i\}$ and $U_i \cup \{v_i\}$ are edges in $\mathcal{H}$. Then, for a fixed $i = 2, \ldots, k$ we call such a choice U_i good, motivated by $W_k = \bigcup_{i \in [k]} U_i$ being an absorbing m-set for T.

Note that in each step $2 \leq i \leq k$ there are $k + (i-1)(k-1) \leq k^2$ vertices in W_{i-1}, thus the number of edges intersecting u_i (or w_i respectively) and at least one other vertex in W_{i-1} is at most $k^2\binom{n}{k-2}$. So the restriction on the minimum degree implies that for each $i \in \{2, \ldots, k\}$ there are at least $2\gamma\binom{n}{k-1} - 2k^2\binom{n}{k-2} \geq \gamma\binom{n}{k-1}$ choices for U_i and in total we obtain $\gamma^{k-1}\binom{n}{k-1}^k/2$ absorbing m-sets for T. □

Continuing the proof of the Lemma, let $\mathcal{L}(T)$ denote the family of all those m-sets absorbing T. From Claim 1 we know $|\mathcal{L}(T)| \geq \gamma^{k-1}\binom{n}{k-1}^k/2$.

Now, choose a family $\mathcal{F}$ of m-sets by selecting each of the $\binom{n}{m}$ possible m-sets independently with probability

$$p = \gamma^k n/\Delta \quad \text{with} \quad \Delta = 2\binom{n}{k-1}^k \geq 2n\binom{n}{m-1} \geq 2m\binom{n}{m}. \tag{3}$$

Then, by Chernoff's bound, with probability $1 - o(1)$, as $n \to \infty$ the family $\mathcal{F}$ fulfills the following properties:

$$|\mathcal{F}| \leq \gamma^k n/m \tag{4}$$

and

$$|\mathcal{L}(T) \cap \mathcal{F}| \geq \gamma^{2k-1} n/5 \quad \forall T \in \binom{V}{k}. \tag{5}$$

Furthermore, using (3) we can bound the expected number of intersecting m-sets by

$$\binom{n}{m} \times m \times \binom{n}{m-1} \times p^2 \leq \gamma^{2k} n/4.$$

Thus, using Markov's bound, we derive that with probability at least $3/4$

$$\mathcal{F} \text{ contains at most } \gamma^{2k} n \text{ intersecting pairs.} \tag{6}$$

Hence, with positive probability the family $\mathcal{F}$ has all the properties stated in (4), (5) and (6). By deleting all the intersecting and non-absorbing m-sets in such a family $\mathcal{F}$ we get a subfamily $\mathcal{F}'$ consisting of pairwise disjoint absorbing m-sets which, due to $\gamma \leq 1/10$, satisfies

$$|\mathcal{L}(T) \cap \mathcal{F}'| \geq \gamma^{2k-1} n/5 - \gamma^{2k} n \geq \gamma^{2k} n \quad \forall T \in \binom{V}{m}.$$

So, since $\mathcal{F}'$ consists of pairwise disjoint absorbing m-sets, $\mathcal{H}[V(\mathcal{F}')]$ contains a perfect matching M of size at most $\gamma^k n/k$. Further, for any subset $W \subset V \setminus V(M)$ of size $\gamma^{2k} n \geq |W| \in k\mathbb{Z}$ we can partition W into at most $\gamma^{2k} n/k$ sets of size k and successively absorb them using a different absorbing m-set each time. Thus there exists a matching covering exactly the vertices in $V(\mathcal{F}') \cup W$. $\qquad \square$

References

[1] S. ABBASI, GÁBOR N. SÁRKÖZY and E. SZEMERÉDI, *A "dereg-ularized" proof of El-Zahar conjecture for large graphs*, submitted.

[2] N. ALON, P. FRANKL, H. HUANG, V. RÖDL, A. RUCISKI and B. SUDAKOV, *Large matchings in uniform hypergraphs and the conjecture of Erds and Samuels*, J. Combin. Theory Ser. A **119** (2012), 1200–1215.

[3] P. CHAU, L. DEBIASIO and H. A. KIERSTEAD, *Pósa's conjecture for graphs of order at least 2108*, Random Structures Algorithms **39** (2011), 507–525.

[4] K. CORRÁDI and A. HAJNAL, *On the maximal number of independent circuits in a graph*, Acta Math. Acad. Sci. Hungar. **14** (1963), 423–439.

[5] B. CSABA, I. LEVITT, J. NAGY-GYÖRGY and E. SZEMERÉDI, *Tight bounds for embedding bounded degree trees. Fete of combinatorics and computer science*, Bolyai Soc. Math. Stud. **20**, János Bolyai Math. Soc., Budapest, 95–137.

[6] B. CSABA, A. JASHMED and E. SZEMEREDI, *Tight bounds for embedding large degree trees*, submitted.

[7] G. A. DIRAC, *Some theorems of abstract graphs*, Proc. London Math. Soc. **3** (1952), 69–81.

[8] A. HAJNAL and E. SZEMERÉDI, *Proof of a conjecture of P. Erdős*, Combinatorial theory and its applications, II, Proc. Colloq., Balatonfüred, 1969, North-Holland, Amsterdam, 1970, 601–623.

[9] H. HÀN and M. SCHACHT, *Dirac-type results for loose Hamilton cycles in uniform hypergraphs*, J. Combin. Theory Ser. B **100** (2010), 332–346.

[10] H. HÀN, Y. PERSON and M. SCHACHT, *On perfect matchings in uniform hypergraphs with large minimum vertex degree* SIAM J. Discrete Math. **23** (2009), 732–748.

[11] A. JASHMED and E. SZEMERÉDI, *A short proof of the conjecture of Seymour*, submitted.

[12] G. Y. KATONA and H. A. KIERSTEAD, *Hamiltonian chains in hypergraphs*, J. Graph Theory **30** (1999), 205–212.

[13] P. KEEVASH, D. KÜHN, R. MYCROFT and D. OSTHUS, *Loose Hamilton cycles in hypergraphs*, Discrete Math. **311** (2011), 544–559.

[14] I. KHAN, *Perfect Matching in 3-uniform hypergraphs with large vertex degree*, arXiv:1101.5830.

[15] I. KHAN, *Perfect Matchings in 4-uniform hypergraphs*, arXiv: 1101.5675.

[16] J. KOMLÓS, G. N. SÁRKÖZY and E. SZEMERÉDI, *On the square of a Hamiltonian cycle in dense graphs*, Proceedings of the Seventh International Conference on Random Structures and Algorithms (Atlanta, GA, 1995), Random Structures Algorithms **9** (1996), 193–211.

[17] J. KOMLÓS, G. N. SÁRKÖZY and E. SZEMERÉDI, *On the Pósa–Seymour conjecture*, J. Graph Theory **29** (1998), 167–176.

[18] J. KOMLÓS, G. N. SÁRKÖZY and E. SZEMERÉDI, *Spanning trees in dense graphs*, Combin. Probab. Comput. **10** (2001), 397–416.

[19] J. KOMLÓS, G. N. SÁRKÖZY and E. SZEMERÉDI, *Proof of a packing conjecture of Bollobás*, Combin. Probab. Comput. **4** (1995), 241–255.

[20] D. KÜHN, R. MYCROFT and D. OSTHUS, *Hamilton l-cycles in k-graphs*, J. Combin. Theory Ser. A **117** (2010), 910–927.

[21] D. KÜHN and D. OSTHUS, *Loose Hamilton cycles in 3-uniform hypergraphs of large minimum degree*, J. Combin. Theory Ser. B **96** (2006), 767–821.

[22] D. KÜHN, D. OSTHUS and A. TREGLOWN, *Matchings in 3-uniform hypergraphs*, J. Combin. Theory Ser. B **103** (2013), 291-305

[23] A. LO and K. MARKSTRÖM, *Minimum codegree threshold for $(K_4^3 - e)$-factors*, J. Combin. Theory Ser. A **120** (2013), 708–721.

[24] V. RÖDL, A. RUCIŃSKI and E. SZEMERÉDI, *A Dirac-type theorem for 3-uniform hypergraphs*, Combin. Probab. Comput. **15** (2006), 229–251.

[25] V. RÖDL, A. RUCIŃSKI and E. SZEMERÉDI, *An approximate Dirac-type theorem for k-uniform hypergraphs*, Combinatorica **28** (2008), 229–260.

[26] V. RÖDL, A. RUCIŃSKI and E. SZEMERÉDI, *Dirac-type conditions for Hamiltonian paths and cycles in 3-uniform hypergraphs*, Adv. Math. **227** (2011), 1225–1299.

Moduli of algebraic and tropical curves

Dan Abramovich

Abstract. *Moduli spaces* are a geometer's obsession. A celebrated example in algebraic geometry is the space $M_{g,n}$ of stable n-pointed algebraic curves of genus g, due to Deligne–Mumford and Knudsen. It has a delightful combinatorial structure based on *weighted graphs*.

Recent papers of Branetti, Melo, Viviani and of Caporaso defined an entirely different moduli space of *tropical curves*, which are weighted metrized graphs. It also has a delightful combinatorial structure based on weighted graphs.

One is led to ask whether there is a geometric connection between these moduli spaces. In joint work [1] with Caporaso and Payne, we exhibit a connection, which passes through a third type of geometry - nonarchimedean analytic geometry.

1. The moduli bug

Geometers of all kinds are excited, one may say obsessed, with moduli spaces; these are the spaces which serve as parameter spaces for the basic spaces geometers are most interested in.

It was none other than Riemann who introduced the moduli bug into geometry, when he noted that Riemann surfaces of genus $g > 1$ "depend on $3g - 3$ *Moduln*". This is a consequence of his famous "Riemann existence theorem", which tells us how to put together a Riemann surfaces by slitting a number of copies of the Riemann sphere and sewing them together. The number $3g - 3$ is simply the number of "effective complex parameters" necessary for obtaining *every* Riemann surface this way. [1]

This brings us to the classic case of the moduli phenomenon, and the obsession that comes with it, the space $\mathcal{M}_g$ of Riemann surfaces of genus g: fixing a compact oriented surface S, each point on $\mathcal{M}_g$ corresponds uniquely to a complex structure C on S. Being an algebraic geometer, I tend to think about these Riemann surfaces as "smooth projective and

Research of Abramovich supported in part by NSF grant DMS-0901278.

[1] It was none other than Riemann who at the same stroke introduced the word "moduli", with which we have been stuck ever since.

connected complex algebraic curves of genus g", or just "curves of genus g" in short. This explains the choice of the letter C.[2]

The moduli space $\mathcal{M}_g$ is the result of important work of many mathematicians, such as Ahlfors, Teichmüller, Bers, Mumford ...

ACKNOWLEDGEMENTS. All results claimed here were obtained jointly with L. Caporaso and S. Payne. Thanks are due to Caporaso and Payne, as well as A. Vistoli and W. Veys who pointed out errors in earlier versions of this text.

2. From elliptic curves to higher genus

The first example, of elliptic curves, is familiar from Complex Analysis, where an elliptic curve is defined as the quotient $\mathbb{C}/\langle 1, \tau \rangle$ of the complex plane by a lattice of rank 2 with $Im(\tau) > 0$. We learn, for instance in Ahlfors's book [2], that isomorphism classes of elliptic curves are identified uniquely by the so called j-invariant $j(\tau)$, an important but complicated analytic function on the upper half plane. In algebraic geometry one can use a shortcut to circumvent this: every elliptic curve has a so called Weierstrass equation

$$E_{a,b}: \ y^2 = x^3 + ax + b$$

with nonzero discriminant $\Delta(a, b) = 4a^3 + 27b^2 \neq 0$. One can identify the j-invariant as

$$j(a, b) = \frac{4a^3}{4a^3 + 27b^2} \in \mathbb{C},$$

so that two elliptic curves are isomorphic: $E_{a,b} \simeq E_{a',b'}$ if and only if $j(a, b) = j(a', b')$.

Either way, the moduli space of elliptic curves is just $\mathbb{C}$ - or, in the language of algebraic geometers, the affine line $\mathbb{A}^1_{\mathbb{C}}$ (Figure 1).

The story for genus $g > 1$ is quite a bit more involved. But the principle, at least in algebraic geometry, is similar: note that $j(a, b)$ is an invariant rational function in the parameters a, b, which are the coefficients of the defining equation of $E_{a,b}$ written in its Weierstrass form. For higher genus one does the same: one finds a sort of canonical form

---

[2] I hope I can be excused for the confusion between "surfaces" and "curves", which comes from the fact that the dimension of $\mathbb{C}$ as a real manifold is 2.

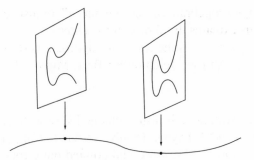

Figure 1. The family of elliptic curves over $\mathbb{C}$.

for a Riemann surface in a suitable projective space, one collects the co-efficients of the defining equations, and then the coordinates on M_g are invariant rational functions in these. The result, in its algebraic version due to Mumford, is:

Theorem 2.1. *The space M_g is a complex quasi projective variety.*

It is a rather nice variety - it is not quite a manifold, but it is an *orbifold*: it locally looks like the quotient of a manifold by the action of a finite group.

In general the global geometry of $\mathcal{M}_g$ is quite a bit more involved than the geometry of $\mathbb{C}$. Its complex dimension is indeed $3g - 3$.

3. The problem of compactness

Angelo Vistoli from Pisa has said that "working with a noncompact space is like trying to keep your change when you have holes in your pockets". The space $\mathbb{C}$ of elliptic curve, and the space $\mathcal{M}_g$ of curves of genus g, are noncompact, and one wishes to find a natural compactification.

Of course every quasi-projective variety sits inside a projective space, and its closure is a compactification. But that is not natural: we want a compactification which is itself a moduli space, of slightly singular Riemann surfaces!

For instance, the moduli space of elliptic curves $\mathbb{C}$ has a nice compactification $\mathbb{P}^1_{\mathbb{C}}$, the Riemann sphere. In which way does the added point ∞ represent a singular Riemann surface?

The function $j(a, b)$ extends to a regular function $j : \mathbb{C}^2 \setminus \{(0, 0)\} \to \mathbb{P}^1$, and indeed all the points where $\Delta = 0$ in $\mathbb{C}^2 \setminus \{(0, 0)\}$ correspond to "singular elliptic curves" such as $E_{-3,2} : y^2 = x^3 - 3x + 2$: the singular point $(x, y) = (1, 0)$ has local plane coordinates with equation of the form $zw = 0$, a so called *node*. In fact all these singular elliptic curves are isomorphic! (See Figure 2.)

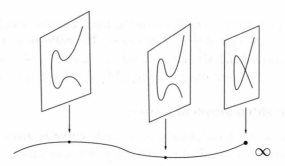

Figure 2. The family of elliptic curves over $\mathbb{P}^1_\mathbb{C}$.

From the Riemann surface point of view one can describe such singular Riemann surfaces as follows: choose your favorite elliptic curve E, thought of as a torus, and wrap a loop around its girth. Now pull the loop until it pops. What you get is the Riemann sphere with two points glued together (Figure 3).

Figure 3. A degenerate elliptic curve as a sphere with glued points.

Deligne and Mumford showed that this can be done in general: they described degenerate algebraic curves of genus g, obtained by choosing a number of disjoint loops and pulling them until they pop (Figure 4).

Figure 4. A degenerate Riemann surface of genus 2.

The result is a singular Riemann surface obtained by taking a number of usual Riemann surfaces, choosing a number of points of them, and indicating how these points are to be glued together (Figure 5).

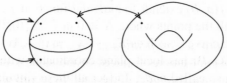

Figure 5. Gluing the same degenerate Riemann surface of genus 2.

One imposes a *stability condition* on the loops chosen, which is best described in combinatorial terms, see below. The collection of all of these objects is Deligne and Mumford's moduli space of *stable curves* $\overline{\mathcal{M}}_g$. It is also a *projective* orbifold containing $\mathcal{M}_g$ as a dense open subset.

4. The weighted graph of a curve

The combinatorial underpinning of a stable curve is given by its *dual graph* Γ. This is a so called *weighted graph* where each vertex v is assigned an integer weight $g(v) \geq 0$.

Given a singular Riemann surface C as above, its graph has a vertex v_i corresponding to each component C_i, weighted by the genus $g(C_i)$. Corresponding to each node, where a point of C_i is glued to a point of C_j one writes an edge tying v_i to v_j (Figure 6).

Figure 6. The glued curve ... and its graph.

The genus of Γ, and of any corresponding singular Riemann surface, is given by a simple formula involving the first Betti number of the graph:

$$b_1(\Gamma, \mathbb{Z}) + \sum_{v \in V(\Gamma)} g(v).$$

We can now describe the stability condition: the graph Γ, and any corresponding curve, is *stable* if every vertex v of genus $g(v) = 0$ has valence $\mathrm{val}(v) \geq 3$ and every vertex v of genus $g(v) = 1$ has valence $\mathrm{val}(v) \geq 1$.

5. The combinatorial structure of moduli space

These graphs give us a way to put together the space $\overline{\mathcal{M}}_g$ piece by piece. For each weighted graph Γ there is a nice moduli space $\mathcal{M}_\Gamma$ parametrizing singular Riemann surfaces with weighted graph Γ. Each $\mathcal{M}_\Gamma$ is an orbifold, and its codimension in $\overline{\mathcal{M}}_g$ is simply the number of edges $|E(\Gamma)|$. We have

$$\overline{\mathcal{M}}_g = \coprod_{g(\Gamma)=g} \mathcal{M}_\Gamma.$$

The pieces $\mathcal{M}_\Gamma$ form a *stratification* of $\overline{\mathcal{M}}_g$, in the sense that the closure of $\mathcal{M}_\Gamma$ is the disjoint union of pieces of the same kind. To determine the combinatorial structure of $\overline{\mathcal{M}}_g$ we need to record which pieces $\mathcal{M}_{\Gamma'}$ appear in the closure: these correspond to singular Riemann surfaces C'

where *more* loops were pulled until they popped than in a curve C corresponding to Γ (Figure 7).

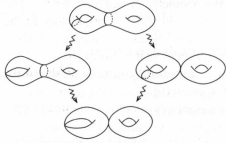

Figure 7. Step-by-step degenerations in genus 2.

On the level of graphs this corresponds to simply saying that there is a *contraction* $\Gamma' \to \Gamma$: on weighted graphs, contracting an edge connecting vertices v_1, v_2 with genera g_1, g_2 results in a vertex with genus g_1+g_2; similarly contracting a loop on a vertex with genus g results with a vertex with genus $g + 1$ (Figure 8).

Figure 8. Contracting an edge ... and a loop.

So the combinatorial structure of $\overline{\mathcal{M}}_g$ is given by the following rule:

$$\mathcal{M}_{\Gamma'} \subset \overline{\mathcal{M}}_\Gamma \iff \exists \text{ contraction } \Gamma' \to \Gamma.$$

The skeletal picture of $\overline{\mathcal{M}}_2$ is given in Figure 9. The top line is the big stratum $\mathcal{M}_2$ of complex dimension 3, and the bottom strata are points, of dimension 0.

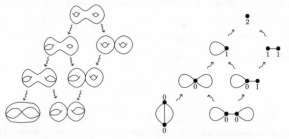

Figure 9. Curves in $\overline{\mathcal{M}}_2$ ... and their graphs.

6. Tropical curves

There is another geometry which builds on the combinatorics of weighted graphs, namely the geometry of *tropical curves*. This is a much more recent theory. One can identify its pre-history in the work of Culler–Vogtmann on outer space [7], and continuing with the work of Mikhalkin on tropical enumeration of plane curves [9, 10]. The notion of tropical curves in the sense described here was introduced by Brannetti–Melo–Viviani [4] and Caporaso [6]. [3]

A *tropical curve* is simply a *metric* weighted graph

$$G = (\Gamma, \ell : E(\Gamma) \to \mathbf{R}_{>0} \cup \{\infty\}).$$

It consists of a weighted graph Γ and a possibly infinite length $\ell(e) > 0$ assigned to each edge.

7. Moduli of tropical curves

Being a geometer, one is infected with the moduli bug. Therefore the moment one writes the definition of a tropical curve one realizes that they have a moduli space. Fixing a weighted graph Γ, the tropical curves having graph isomorphic to Γ are determined by the lengths of the edges, and the collection of lengths is unique up to the permutations obtained by automorphisms of the graph. We can therefore declare the moduli space of such tropical curves to be

$$\mathcal{M}_\Gamma^{\mathrm{Trop}} = (\mathbf{R}_{>0} \cup \{\infty\})^{|E(\Gamma)|} / \mathrm{Aut}(\Gamma).$$

We can put together these moduli spaces $\mathcal{M}_\Gamma^{\mathrm{Trop}}$ by observing the following: if we let the length of an edge e in G approach 0, the metric space G approaches G', which is the metric graph associated to the graph Γ' obtained by contracting e, as in Figure 10.

Figure 10. Pulling an edge ... and a loop.

---

[3] The name "tropical" is the result of tradition: tropical curves arise in tropical geometry, which sometimes relies on min-plus algebra. This was studied by the Brazilian (thus tropical) computer scientist Imre Simon.

This allows us to glue together $\mathcal{M}_\Gamma^{\mathrm{Trop}}$ in one big moduli space

$$\overline{\mathcal{M}}_g^{\mathrm{Trop}} = \coprod_{g(\Gamma)=g} \mathcal{M}_\Gamma^{\mathrm{Trop}}.$$

It is a nice compact cell complex.

Note that the gluing rule precisely means that

$$\overline{\mathcal{M}_{\Gamma'}^{\mathrm{Trop}}} \supset \mathcal{M}_\Gamma^{\mathrm{Trop}} \quad \Longleftrightarrow \quad \exists \text{ contraction } \Gamma' \to \Gamma.$$

8. The question of comparison

We obtained two geometries associated to the combinatorics of weighted graphs, summarized as follows:

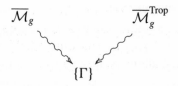

The moduli spaces are clearly similar in their combinatorial nature, and one would like to tie them together somehow:

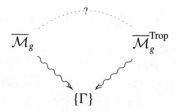

The question was raised in [6, Section 6]. A possible answer was suggested by Baker-Payne-Rabinoff [5, Remark 5.52], Tyomkin [12, Section 2], Viviani [13, Theorem A]. Below I report on joint work [1] with Caporaso and Payne, where we prove the suggested answer to be correct; it is based on the non-archimedean analytic spaces of Berkovich [3] and their skeleta, and fundamental constructions of such skeleta by Thuillier [11].

One slightly disturbing feature is the fact that the combinatorial structures - the stratifications - of the moduli spaces $\overline{\mathcal{M}}_g$ and $\overline{\mathcal{M}}_g^{\mathrm{Trop}}$ are reversed! In Figure 11, the top stratum is a point in dimension 0, and the bottom strata are of dimension 3.

dimension

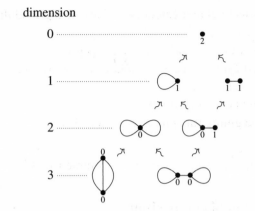

Figure 11. Graph contractions in genus 2.

9. Nonarchimedean analytic geometry

A *valued field* is a field K with a multiplicative seminorm $K \to \mathbf{R}_{\geq 0}$. One often translates the seminorm to a *valuation* $v : K \to \mathbf{R} \cup \{\infty\}$ by declaring $\mathrm{val}(x) = -\log \|x\|$. The valued field is *nonarchimedean* if, like the field of p-adic numbers, it satisfies the strict triangle inequality

$$\|a + b\| \leq \max(\|a\|, \|b\|).$$

One defines the valuation ring to be

$$R = \{x \in K : \|x\| \leq 1\} = \{x \in K : \mathrm{val}(x) \geq 0\},$$

which has a unique maximal ideal $I = \{x \in K : \|x\| < 1\}$. The residue field is defined as $\kappa = R/I$.

Recall that in scheme theory, a point of a variety X corresponds to a field extension $K \supset \mathbb{C}$ and a morphism $\mathrm{Spec}\, K \to X$, up to equivalence given by further extensions. In particular a point of $\overline{\mathcal{M}}_g$ corresponds to a field extension $K \supset \mathbb{C}$ and a stable curve $C/\mathrm{Spec}\, K$. Topologists cannot be happy about this structure, because the Zariski topology of a scheme is not Hausdorff in the least.

Berkovich associates to X an analytic variety X^{An} - a locally ringed space which admits a natural morphism $X^{\mathrm{An}} \to X$. A point of X^{An} corresponds to a nonarchimedean valued field extension $K \supset \mathbb{C}$ (extending the *trivial* valuation on $\mathbb{C}$) and a morphism $\mathrm{Spec}\, K \to X$, up to equivalence by further valued field extensions.

In particular a point of $\overline{\mathcal{M}}_g^{\mathrm{An}}$ corresponds to a *valued* field extension $K \supset \mathbb{C}$ and a stable curve $C/\mathrm{Spec}\, K$.

Since every valued field extension is, in particular, a field extension, there is a morphism $\overline{\mathcal{M}}_g^{\text{An}} \to \overline{\mathcal{M}}_g$. It is a bit of magic that adding the valuations "stretches" the generic points just enough to make the space Hausdorff and locally connected. We can now extend our diagram of relationships as follows:

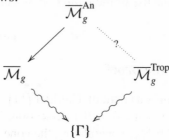

10. Making the connection

We are finally ready to close the diagram.

We said that a point of $\overline{\mathcal{M}}_g^{\text{An}}$ corresponds to a valued field extension $K \supset \mathbb{C}$ and a stable curve $C/\operatorname{Spec} K$. Since $\overline{\mathcal{M}}_g$ is proper, a stable curve over K uniquely extends to $C/\operatorname{Spec} R$ over the valuation ring (at least after a further field extension, which we may ignore because of our equivalence relation). The fiber of $C/\operatorname{Spec} R$ over the residue field is a stable curve C_s with dual graph $\Gamma(C_s)$. We need to put a metric on this graph.

Consider an edge $s \in E(\Gamma(C_s))$ corresponding to a node $p \in C_s$. Near p the curve C admits a local equation of the shape $xy = f$, where $f \in R$. Define $\ell(e) = \operatorname{val}(f)$, which is independent of the local equation chosen (or the field K). This results with a tropical curve $G := (\Gamma, \ell)$.

Theorem 10.1 ([1]). *The resulting map* $\operatorname{Trop} : \overline{\mathcal{M}}_g^{\text{An}} \to \overline{\mathcal{M}}_g^{\text{Trop}}$ *is proper, continuous and surjective. It also makes* $\overline{\mathcal{M}}_g^{\text{Trop}}$ *canonically isomorphic to the Berkovich skeleton of* $\overline{\mathcal{M}}_g^{\text{An}}$.

We obtain a picture as follows:

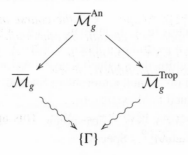

In other words, we tied together the algebraic geometry of the space $\overline{\mathcal{M}}_g$ of stable curves with the metric geometry of the space $\overline{\mathcal{M}}_g^{\mathrm{Trop}}$ of tropical curves, by going through nonarchimedean analytic geometry. What enabled us to relate the spaces $\overline{\mathcal{M}}_g$ and $\overline{\mathcal{M}}_g^{\mathrm{Trop}}$ with "reversed" combinatorial structures was the use of the *reduction* of a curve over K to a curve over the residue field.

11. Comments on the proof

A key tool in our proof is a result of Thuillier [11]. To a complex toroidal embedding X, with or without self intersection, Thuillier assigns a so called *compactified fan* Σ_X, generalizing the cone complex construction of [8]. This is a canonical construction of a Berkovich skeleton in the special case of toroidal varieties over a field with trivial valuation (such as $\mathbb{C}$). The combinatorial structure of the fan mirrors that of the toroidal structure: there is a cone $\sigma_F \in \Sigma_X$ assigned to each stratum $F \in X$, and these are glued to each other via a natural rule which in particular says that σ_F is glued as a face of $\sigma_{F'}$ precisely if $F' \subset \overline{F}$ - note the reversal of order!

Furthermore, considering the associated restricted Berkovich space $X^{\beth}$, one has that Σ_X canonically sits inside $X^{\beth}$ as a deformation retract, in particular there is a canonical proper continuous map $p : X^{\beth} \to \Sigma_X$. This map is described via reduction: if K is a valued field and $x :$ Spec $K \to X$ is a point on $X^{\beth}$, consider the reduction $\bar{x} :$ Spec $\kappa \to X$, and assume $\bar{x}$ lands in the toroidal stratum F. Then $p(x)$ lands in the cone σ_F corresponding to the stratum F. The position of $p(x)$ on σ_F is determined by by applying the toroidal valuations at F to x.

This does not quite apply to the moduli space, but almost. The point is that the moduli space is not toroidal, but the moduli stack is. The technical point we needed to prove is the following slight generalization of Thuillier's result:

Proposition 11.1 ([1]). *Suppose X is the coarse moduli space of a toroidal stack. Then there is a canonical compactified fan Σ_X which is a deformation retract of $X^{\beth}$. The map $p : X^{\beth} \to \Sigma_X$ admits the same description in terms of reductions as above.*

When X is proper we have $X^{\beth} = X^{\mathrm{An}}$. This applies to the moduli space $\overline{\mathcal{M}}_g$, so $\overline{\mathcal{M}}_g^{\beth} = \overline{\mathcal{M}}_g^{\mathrm{An}}$.

We obtain the following picture

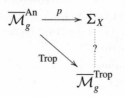

There is also an observation to be made. The fan $\Sigma_{\overline{\mathcal{M}}_g}$ is put together based on the strata of $\overline{\mathcal{M}}_g$, namely the spaces $\overline{\mathcal{M}}_\Gamma$. The corresponding cones $\sigma_{\overline{\mathcal{M}}_\Gamma}$ are glued together via the same rule used to glue together the space $\overline{\mathcal{M}}_g^{\text{Trop}}$. Putting these facts together leads to the following:

Theorem 11.2 ([1]). *We have a canonical isomorphism* $\Sigma_X \simeq \overline{\mathcal{M}}_g^{\text{Trop}}$, *making the following diagram commutative:*

$$\overline{\mathcal{M}}_g^{\text{An}} \xrightarrow{\ p\ } \Sigma_X$$

$$\overset{\text{Trop}}{\searrow} \quad \Vert$$

$$\overline{\mathcal{M}}_g^{\text{Trop}}$$

In particular the map Trop is proper and continuous, as claimed.

References

[1] D. ABRAMOVICH, L. CAPORASO and S. PAYNE, *The tropicalization of moduli space*, http://arxiv.org/abs/1212.0373

[2] LARS V. AHLFORS, "Complex Analysis", third ed., McGraw-Hill Book Co., New York, 1978, *An introduction to the theory of analytic functions of one complex variable*, International Series in Pure and Applied Mathematics.

[3] V. BERKOVICH, "Spectral Theory and Analytic Geometry over non-Archimedean Fields", Mathematical Surveys and Monographs, Vol. 33, American Mathematical Society, Providence, RI, 1990.

[4] S. BRANNETTI, M. MELO and F. VIVIANI, *On the tropical Torelli map*, Adv. Math. **226** (2011), 2546–2586.

[5] M. BAKER, S. PAYNE and J. RABINOFF, *Nonarchimedean geometry, tropicalization, and metrics on curves*, ArXiv e-prints (2011), arXiv:1104.0320.

[6] L. CAPORASO, *Algebraic and tropical curves: comparing their moduli spaces*, 2011, http://arxiv.org/abs/1101.4821.

[7] M. CULLER and K. VOGTMANN, *Moduli of graphs and automorphisms of free groups*, Invent. Math. **84** (1986), 91–119.

[8] G. KEMPF, F. F. KNUDSEN, D. MUMFORD and B. SAINT-DONAT, "Toroidal Embeddings", I, Lecture Notes in Mathematics, Vol. 339, Springer-Verlag, Berlin, 1973.

[9] G. MIKHALKIN, *Enumerative tropical algebraic geometry in* $\mathbb{R}^2$, J. Amer. Math. Soc. **18** (2005), 313–377.

[10] G. MIKHALKIN, *Tropical geometry and its applications*, International Congress of Mathematicians, Vol. II, Eur. Math. Soc., Zürich, 2006, 827–852.

[11] A. THUILLIER, *Géométrie toroïdale et géométrie analytique non archimédienne. Application au type d'homotopie de certains schémas formels*, Manuscripta Math. **123** (2007), 381–451.

[12] I. TYOMKIN, *Tropical geometry and correspondence theorems via toric stacks*, ArXiv e-prints (2010), arXiv:1001.1554.

[13] F. VIVIANI, *Tropicalizing vs Compactifying the Torelli morphism*, ArXiv e-prints (2012), arXiv:1204.3875.

Three projection operators in complex analysis

Elias M. Stein

1. Introduction

This lecture reports on joint work with Loredana Lanzani concerned with three types of projection operators arising in the setting where Ω is an appropriate bounded domain in $\mathbb{C}^n$. One paper containing the proof of some of the results stated here is available in [8]; several others are being prepared for publication.

The first kind of operator we will deal with is represented by the Cauchy integral when $n = 1$, and variants of the Cauchy-Fantappié integral when $n > 1$. The second type is the Cauchy-Szegö projection associated to $L^2(\partial\Omega)$, and the third is the Bergman projection associated to $L^2(\Omega)$. All three can be viewed as projection operators: the first two on the subspace of $L^2(\partial\Omega)$ of restrictions of holomorphic functions, and the third on the subspace of holomorphic functions belonging to $L^2(\Omega)$. The problem we are concerned with, and to which we give a partial answer, can be stated succinctly as follows:

What are the minimal smoothness conditions on the boundary of Ω that guarantee that these operators are bounded on L^p?

2. The case $n=1$

This problem has a long and illustrious history in the case of one complex dimension. Let us review it briefly. (Some details can be found in [2,5,7], which contain further citations to the literature.)

* Supported by a National Science Foundation award DMS-0901040.

Suppose Ω is a bounded domain in $\mathbb{C}$ whose boundary $\partial\Omega$ is a rectifiable curve γ. Then the Cauchy integral is given by

$$C(f)(z) = \frac{1}{2\pi i} \int_{\partial\Omega} \frac{f(w)dw}{w - z} , \quad \text{for } z \in \Omega.$$

When Ω is the unit disc, then a classical theorem of M. Riesz says that the mapping $f \mapsto C(F)|_{\partial\Omega}$, defined initially for f that are (say) smooth, is extendable to a bounded operator on $L^p(\partial\Omega)$, for $1 < p < \infty$. Very much the same result holds when γ, the boundary of Ω, is of class $C^{1+\epsilon}$, with $\epsilon > 0$, (proved either by approximating to the result when γ is the unit circle, or adapting one of the several methods of proof used in the classical case). However in the limiting case when $\epsilon = 0$, these ideas break down and new methods are needed. The theorems proved by Calderón, Coifman, McIntosh, Meyers and David (between 1977-1984) showed that the corresponding L^p result held in the following list of increasing generality: the boundary is of class C^1; it is Lipschitz (the first derivatives are merely bounded and not necessarily continuous); and the boundary is an "Ahlfors-regular" curve.

We pass next to the Cauchy-Szegö projection S, the corresponding orthogonal projection with respect to the Hilbert space structure of $L^2(\partial\Omega)$. In fact when Ω is the unit disc, the two operators C and S are identical. Let us now restrict our attention to the case when Ω is simply connected and when its boundary is Lipschitz. Here a key tool is the conformal map $\Phi : D \to \Omega$, where D is the unit disc. We consider the induced correspondence τ given by $\tau(f)(e^{i\theta}) = (\Phi'(e^{i\theta}))^{\frac{1}{2}} f(\Phi(e^{i\theta}))$, and the fact that $S = \tau^{-1} S_0 \tau$, where S_0 is the Szegö projection for the disc D. Using ideas of Calderón, Kenig, Pommerenke and others, one can show that $|\Phi'|^r$ belongs to the Muckenhaupt class A_p, with $r = 1 - p/2$, from which one gets the following. As a consequence, if we suppose that $\partial\Omega$ has a Lipschitz bound M, then S is bounded on L^p,

- For $p'_M < p < p_M$. Here p_M depends on M, but $p_M > 4$.
- For $1 < p < \infty$, if $\partial\Omega$ is of class C^1.

There is an alternative approach to the second result that relates the Cauchy-Szegö projection S to the Cauchy integral C. It is based on the following identity, used in [6]

$$S(I - D) = C , \quad \text{where } D = C^* - C. \tag{2.1}$$

There are somewhat analogous results for the Bergman projection in the case of one complex dimension. We shall not discuss this further, but refer the reader to the papers cited above.

3. Cauchy integral in $\mathbb{C}^n$, $n > 1$; some generalities

We shall see that a very different situation occurs when trying to extend the results of Section 2 to higher dimensions. Here are some new issues that arise when $n > 1$:

1. There is no "God given" (unique) Cauchy integral associated to a domain Ω.
2. Pseudo-convexity of Ω, must, in one form or another, play a role.
3. Since this condition involves (implicitly) two derivatives, the "best" results are to be expected "near" C^2, (as opposed to near C^1 when $n = 1$).

In view of the non-uniqueness of the Cauchy integral (and its problematic existence), it might be worthwhile to set down the minimum conditions that would be required of candidates for the Cauchy integral. We would want such an operator C given in the form

$$C(f)(z) = \int_{\partial\Omega} K(z, w) f(w) d\sigma(w), \ z \in \Omega, \tag{3.1}$$

to satisfy:

(a) The kernel K should be given by a "natural" or explicit formula (at least up to first approximation) that involves Ω.
(b) The mapping $f \mapsto C(f)$ reproduces holomorphic functions. In particular if f is continuous in $\overline{\Omega}$ and holomorphic in Ω then $C(f)(z) = f(z)$, for $z \in \Omega$.
(c) $C(f)(z)$ is holomorphic in $z \in \Omega$, for any f that is given and continuous on $\partial\Omega$.

Now there is a formalism (the Cauchy-Fantappié formalism of Fantappié (1943), Leray, and Koppleman (1967)), which provides Cauchy integrals satisfying the requirements (a) and (b) in a general setting. Condition (c) however, is more problematic, even when the domain is smooth. Constructing such Cauchy integrals has been carried out only in particular situations, two of which are discussed below. (A detailed exposition of some of these questions can be found in [9, Chapter IV].)

The Cauchy-Fantappié construction proceeds once one is given a generating form

$$G(z, w) = \sum_{k=1}^{n} g_k(z, w) dw_k$$

with the property that if

$$\Delta(z, w) = \langle G, w - z \rangle = \sum_{k=1}^{n} g_k(z, w)(w_k - z_k), \quad \text{then}$$

$$\Delta(z, w) \neq 0 \quad \text{if } z \in \Omega \text{ and } w \in \partial\Omega. \tag{3.2}$$

With such a form, the corresponding Cauchy integral C is defined by

$$C(f)(z) = \frac{1}{(\partial \pi i)^n} \int\limits_{\partial\Omega} \frac{G \wedge (\bar{\partial}G)^{n-1}}{\Delta^n} f(w), \tag{3.3}$$

where $(\bar{\partial}G)^{n-1} = (\bar{\partial}_w G) \wedge (\bar{\partial}_w G) \wedge \cdots$ ($n - 1$ wedge products), and this C satisfies conditions (a) and (b).

There are many choices of the generating form G that assure (a) and (b). However, the condition (c) which comes about if G is analytic in z, is much harder to come to grips with, and has been dealt with in two circumstances, when Ω is smooth:

(i) If Ω is convex. Then one can take, with Leray, $G = \partial\rho$, where ρ is a defining function, so $\Delta(z, w) = \sum_{k=1}^{n} \dfrac{\partial\rho}{\partial w_k}(w)(w_k - z_k)$ and the convexity of Ω assures that

$$\Delta(z, w) \neq 0, \quad \text{for } z \in \Omega \text{ and } w \in \partial\Omega.$$

(ii) If Ω is strongly pseudo-convex, there are constructions of Henkin and Ramirez for appropriate generating forms G based on the Levi polynomial (in z) centered at $w \in \partial\Omega$ (see also below). There is a also a later construction in [6] that would be relevant for us.

4. Statement of the first result: the Cauchy-Leray integral

We now come to the results recently obtained jointly with L. Lanzani. Here we do not assume Ω is smooth, but only that it satisfies a condition near C^2, as indicated above. More precisely, we shall assume that Ω is of class $C^{1,1}$ (that it has a defining function ρ whose first derivatives are Lipschitz functions). The basic geometric assumption we make in this case is that Ω is *strongly $\mathbb{C}$-linearly convex* in the following sense:

$$d(z, T_w^{\mathbb{C}}) \geq C|z - w|^2, \tag{4.1}$$

where $c > 0$, for every $z \in \Omega$ and $w \in \partial\Omega$. Here $T_w^{\mathbb{C}}$ is the "complex" sub-space of tangent space T_w at w, and d is the Euclidean distance.

To understand condition (4.1) we make the following remarks: If we assumed Ω was of class C^2 (instead of class $C^{1,1}$) then condition (4.1) would imply that

$$\nabla_T^2 \rho \Big|_{T^{\mathbb{C}}} \geq cI \qquad (4.2)$$

and hence that Ω is strongly pseudo-convex.

Also, if Ω were strongly convex it would automatically satisfy (4.1). Finally, in the $C^{1,1}$ case, by arguments mentioned below, one can see that (4.1) implies that (4.2) holds for $d\sigma$ - a.e. w in $\partial\Omega$.

With these preliminaries we can state our first main result.

Theorem 4.1. *Suppose Ω is a bounded domain of class $C^{1,1}$ which is strongly $\mathbb{C}$-linearly convex. Then there is a natural definition of the Cauchy-Leray integral (3.3), (here $G = \partial\rho$), so that the mapping $f \mapsto C(f)$ initially defined for $f \in C'(\partial\Omega)$, extends to a bounded operator on $L^p(\partial\Omega)$, $1 < p < \infty$.*

Note that here L^p is taken with respect to the induced Lebesgue measure $d\sigma$ on $\partial\Omega$.

5. Some ideas related to the proof

First, we explain the main difficulty in defining the Cauchy-Leray integral in the case of $C^{1,1}$ domains. It arises from the fact that the definition (3.3) involves *second* derivatives of the defining function ρ. However ρ is only assumed to be of class $C^{1,1}$, so that these derivatives are L^∞ functions on $\mathbb{C}^n$, and as such not defined on $\partial\Omega$ which has $2n$-dimensional Lebesgue measure zero.

What gets us out of this quandary is that here, in effect, not all second derivatives are involved but only those that are "tangential" to $\partial\Omega$. Matters are made precise by the following "restriction" principle and its variants.

Suppose $F \in C^{1,1}(\mathbb{C}^n)$ and we want to define $\bar{\partial}\partial F \Big|_{\partial\Omega}$. We note that if F were of class C^2 we would have

$$\int_{\partial\Omega} j^*(\bar{\partial}\partial F) \wedge \Psi = -\int_{\partial\Omega} j^*(\partial F) \wedge d\Psi, \qquad (5.1)$$

where Ψ is any $2n - 3$ format class C^1, and here j^* is the induced mapping to forms on $\partial\Omega$.

Proposition 5.1. *For $F \in C^{1,1}(\mathbb{C}^n)$, there exists a unique 2-form $j^*(\bar{\partial}\partial F)$ in $\partial\Omega$ with $L^\infty(d\sigma)$ coefficients so that (4.2) holds.*

This is a consequence of an approximation lemma: There is a sequence $\{F_n\}$ of C^∞ functions on $\mathbb{C}^n$, that are uniformly bounded in the $C^{1,1}(\mathbb{C}^n)$ norm, so that $F_k \to F$ and $\nabla F_k \to \nabla F$ uniformly on $\partial\Omega$, and moreover $\nabla_T^2 F_n$ converges $(d\sigma)$ a.e. on $\partial\Omega$. Here $\nabla_T^2 F$ is the "tangential" restriction of the Hessian $\nabla^2 F$ of F. Moreover the indicated limit, which we may designate as $\nabla_T^2 F$, is independent of the approximating sequence $\{F_n\}$.

A second point of interest is to work with the "Levi-Leray" measure $d\mu_\rho$ which we can define as follows.

We take the linear functional

$$\ell(f) = \frac{1}{(2\pi i)^n} \int_{\partial\Omega} f(w)(\partial\rho) \wedge (\bar{\partial}\partial_\rho)^{n-1}$$

and write $\ell f = \int_{\partial\Omega} f d\mu_\rho$. We then have, by a variant of the proposition, $d\mu_\rho = D(w)d\sigma$, and $D(w) = c \det(\nabla_{T^c}^2 \rho)|\nabla_\rho)|$, and $D(w) \approx 1$, via (4.2) and the calculation in [R, p. 289] in the case ρ is of class C^2.

With this we have that the Cauchy-Leray integral becomes

$$C_L(f)(z) = \int_{\partial\Omega} \frac{f(w)d\mu_\rho(w)}{(\Delta(z,w))^n} \tag{5.2}$$

In studying (5.2) we apply the "T(1)-theorem" in [3], where the underlying geometry is determined by the quasi-metric $|\Delta(w,z)|^{\frac{1}{2}}$. In this metric, the ball centered at w, reacting to z has $d\mu_\rho$-measure $\approx |\Delta(w,z)|^n$.

The study of (5.2) also requires that we verify cancellation properties in terms of its action on "bump functions." These matters again differ from the case $n = 1$, and in fact there is an unexpected favorable twist: the kernel in (5.2) is an appropriate derivative, as can be suspected by the observation that on the Heisenberg group one has $(|z|^2 + it)^{-n} = c' \frac{d}{dt}(|z|^2 + it)^{-n+1}$, if $n > 1$. (However for $n = 1$, the corresponding identity involves the logarithm!). Indeed by an integration-by-parts argument that uses (5.1) we see that when $n > 1$ and f is of class C^1,

$$C_L(f)(z) = c \int_{\partial\Omega} \frac{df(w) \wedge j^*(\bar{\partial}\partial\rho)^{n-1}}{\Delta(z,w)^{n-1}} + E(f)(z),$$

where

$$E(f)(z) = \int_{\partial\Omega} E(z, w) f(w) d\sigma(w),$$

with

$$E(z, w) = 0(|z - w| \, |\Delta(z, w)|^{-n}),$$

so that the operator E is a negligible term.

A final point is that the hypotheses of the theorem are in the nature of best possible. In fact, Barett and Lanzani [1] have given examples of Reinhardt domains where the L^2 result fails when a condition near C^2 is replaced by $C^{2-\epsilon}$, or "strong" pseudo-convexity is replaced by its "weak" analogue.

6. The Szegö and Bergman projections

In studying these two projections we restrict our attention to bounded domains Ω whose boundaries are of class C^2 and which are strongly pseudo-convex. Thus a comparison with the situation for the Cauchy-Leray integral, here the regularity assumption in stronger (C^2 vs. $C^{1,1}$), but the geometric requirement is weaker (strong pseudo-convexity vs. strong C-linear convexity).

For the Cauchy-Szegö projection we take as our underlying space $L^2(\partial\Omega, d\sigma)$, and we define S to be the orthogonal projections onto the closure of the subspace of functions that arise as the restriction to $\partial\Omega$ of holomorphic functions in a neighborhood of $\overline{\Omega}$. Similarly for the Bergman projection, the underlying space is $L^2(\Omega, dV)$, (with dV the Euclidean volume on $\mathbb{C}^n$), and the Bergman projection B is the orthogonal projection on the subspace of holomorphic functions. The following two results were obtained jointly with L. Lanzani.

Theorem 6.1. *Under the assumption that the bounded domain Ω has a C^2 boundary and is strongly pseudo-convex, one can assert that S extends to a bounded mapping on $L^p(\partial\Omega, d\sigma)$, when $1 < p < \infty$.*

Theorem 6.2. *Under the same assumptions on Ω it follows that the operator B extends to a bounded operator on $L^p(\Omega, dV)$ for $1 < p < \infty$.*

The following additional results also hold.

Corollary 6.3. *The result of Theorem 6.1 extends to the case when the projection S is replaced by the corresponding orthogonal projection S_w, with respect to the Hilbert space $L^2(\partial\Omega, wd\sigma)$ where w is any continuous strictly positive function on $\partial\Omega$.*

Also a similar variant of Theorem 6.2 holds for B_w the orthogonal projection on the sub-space of $L^2(\Omega, w \, dV)$. Here w is any strictly positive continuous function on $\overline{\Omega}$.

Corollary 6.4. *One also has the L^p boundedness of the operator $|B|$, whose kernel is $|B(z, w)|$, where $B(z, w)$ is the Bergman kernel.*

We make the following remarks to clarify the background of these results.

1. There is no simple and direct relation between S and S_w, nor between B and B_w. Thus the results for general w are not immediate consequences of the results for $w \equiv 1$.
2. When $\partial\Omega$ and w are smooth (*i.e.* C^k for sufficiently high k), the above results have been known for a long time. In fact if Ω is the unit ball, then the Szegö kernel is $c(1 - z\overline{w})^{-n}$, with $B(z, w) = c'(1 - z\overline{w})^{-n-1}$, for appropriate c and c'. Moreover when $\partial\Omega$ and w are smooth (and $\partial\Omega$ is strongly pseudo-convex), there are analogous asymptotic formulas for the kernels in question due to [4], which allow a proof of Theorems 6.1 and 6.2 in these cases.
3. Another approach to Theorem 6.2 in the case of smooth strongly pseudo-convex domains is via the $\overline{\partial}$-Neumann problem, but we shall not say anything more about this here.

7. Outline of the proofs

There are three main steps in the proof of Theorem 6.1.

  (i) Construction of a Cauchy-Fantappié type integral C_ϵ.
 (ii) Estimates for $C_\epsilon - C_\epsilon^*$
(iii) Application of a variant of identity (2.1).

Step (1). Here one first augments the choice of the $g_k = \frac{\partial\rho}{\partial w_k}(w)$, used for the Cauchy-Leray integral, and takes

$$\widetilde{g}_k(z, w) = \frac{\partial\rho}{\partial w_k}(w) + \frac{1}{2} \sum_{j=1}^{n} \frac{\partial^2\rho}{\partial w_j \partial w_k}(w)(w_j - z_j).$$

So here we consider first $\widetilde{\Delta}(z, w) = \sum_{k=1}^{n} \widetilde{g}_k \cdot (w_k - z_k)$, which is the Levi polynomial (in z) centered at w.

The strong pseudo convexity of Ω (and an appropriate defining function ρ) guarantee that

$$Re(\widetilde{\Delta}(z, w)) \geq |z - w|^2,$$

whenever $z - w$ is sufficiently small.

To proceed further and obtain a Cauchy-Fantappié integral (see the requirement (3.2)) one must modify this construction. There is now the additional problem that $\widetilde{\Delta}(z, w)$ is only continuous in w in our situation. Here we follow the route set out in [6], as adapted in [9], (where as opposed to what follows below, there is only one fixed ϵ).

Because $\left\{\dfrac{\partial^2 \rho(w)}{\partial w_j \partial w_k}\right\}$ is here only continuous, we replace it by a smooth (at least C^1) matrix $\left\{\dfrac{\tau^\epsilon(w)}{\tau_{j,k}^\epsilon(w)}\right\}$ so that $\left|\dfrac{\partial^2 \rho(w)}{\partial w_j \partial w_k} - \tau_{j,k}^\epsilon(w)\right| < \epsilon$ for $w \in \partial\Omega$. As a result we define

$$g_k^\epsilon(z, w) = \chi \left(\frac{\partial \rho(w)}{\partial w_k} + \frac{1}{2} \sum_j \tau_{jk}^\epsilon(w_j - z_j)\right) + (1-\chi)(\overline{w}_k - \overline{z}_k), \quad (7.1)$$

where χ is a smooth cut-off function which is supported in a neighborhood of the diagonal, and $\chi = 1$ in a smaller neighborhood.

Now let C_ϵ^1 be the Cauchy-Fantappié integral (3.3) with generating form $G = \displaystyle\sum_{k=1}^n g_k^\epsilon(z, w) dw_k$. Then it satisfies requirement (a) and (b) of Section 3, but not (c), because the intervention of χ and $\overline{z}_k - \overline{w}_k$ destroys the holomorphicity in z. However C_ϵ^1 can be corrected via a solution of a $\overline{\partial}$ problem, so that $C_\epsilon = C_\epsilon^1 + C_\epsilon^2$ satisfies condition (c), together with (a) and (b). One notes that the correction C_ϵ^2 is "harmless" since its kernel is bounded as (z, w) ranges over $\overline{\Omega} \times \partial\Omega$.

Finally, using a methodology similar to the proof of Theorem 4.1 one shows

$$\|C_\epsilon(f)\|_{L^p} \leq c_{\epsilon,p}\|f\|_{L^p}, \quad 1 < p < \infty.$$

However it is important to point out, that in general the bound $c_{\epsilon,p}$ grows to infinity as $\epsilon \to 0$, so that the C_ϵ can *not* be genuine approximations of S. Nevertheless we shall see below that in a sense the C_ϵ give us critical information about S.

Step (2). Here the goal is the following splitting:

Proposition 7.1. *Given $\epsilon > 0$, there is a $\delta = \delta_\epsilon$ so that we can write*

$$C_\epsilon - C_\epsilon^* = A_\epsilon + B_\epsilon \quad where$$

- $\|A_\epsilon\|_{L^p \to L^p} \leq \epsilon c_p$, $\quad 1 < p < \infty$
- *The operator B_ϵ has a bounded kernel, hence B_ϵ maps $L^1(\partial\Omega)$ to $L^\infty(\partial\Omega)$.*

We note that in fact the bound of the kernel of B_ϵ may grow to infinity as $\epsilon \to 0$.

To prove the proposition we first reverify an important "symmetry" condition:

For each ϵ, there is a δ_ϵ, so that

$$|\Delta_\epsilon(z, w) - \overline{\Delta}_\epsilon(w, z)| \leq \epsilon c |z - w|^2, \tag{7.2}$$

if $|z - w| < \delta_\epsilon$.

Here we have set $\Delta_\epsilon(z, w) = \sum g_k^\epsilon(z, w)(w_k - z_k)$, with the g_k^ϵ given by (7.1).

With this one proceeds as follows. Suppose $H_\epsilon(z, w)$ is the kernel of the operator $C_\epsilon - C_\epsilon^*$. Then we take A_ϵ and B_ϵ to be the operators with kernels respectively $\chi_\delta(z - w)H_\epsilon(z, w)$ and $(1 - \chi_\delta(z - w))H_\epsilon(z, w)$, where $\chi_\delta(z - w) = \chi\left(\frac{z-w}{\delta}\right)$ and $\delta = \delta_\epsilon$, chosen according to (7.2).

Step (3). We conclude the proof of Theorem 6.1 by using an identity similar to (2.1):

$$S(I - (C_\epsilon^* - C_\epsilon)) = C_\epsilon, \tag{7.3}$$

Hence $S(I - A_\epsilon) = C_\epsilon + SB_\epsilon$

Now for each p, take $\epsilon > 0$ so that $\epsilon c_p \leq \frac{1}{2}$.

Then $I - A_\epsilon$ is invertible and we have

$$S = (C_\epsilon + SB_\epsilon)(I - A_\epsilon)^{-1}.$$

Since $(I - A_\epsilon)^{-1}$ is bounded on L^p, and also C_ϵ, it suffices to see that SB_ϵ is also bounded on L^p. Assume too the moment that $p \leq 2$. Then since B_ϵ maps L^1 to L^∞, it maps L^p to L^2, while S maps L^2 to itself, yielding the fact that SB_ϵ is bounded on L^p. The case $2 \leq p$ is obtained by dualizing this argument.

The proof of Theorem 6.2 has a similar outline, but the details are simpler since we are dealing with operators that converge absolutely (as suggested by Corollary 6.4). Thus one can avoid the delicate $T(1)$-theorem machinery and make instead absolutely convergent integral estimates. However to start the process we need an analogue of the operators C_ϵ where the integration is performed over Ω instead of $\partial\Omega$. The Cauchy-Fantappié formalism needed for that is an idea of E. Ligocka, described in [9, Chapter VII, Section 7].

References

[1] D. BARRETT and L. LANZANI, *The spectrum of the Leray transform for convex Reinhardt domains in C^2*, J. Funct. Anal. **257** (2009), 2780–2819.

[2] G. DAVID, *Opérateurs intégraux singuliers sur certain courbes du plan complexe*, Ann. Sci. École Norm. Sup. **17** (1984), 157–189.

[3] G. DAVID, J. L. JOURNÉ and S. SEMMES, *Opérateurs de Calderón-Zygmund, functions para-acrétives et interpolation*, Rev. Mat. Iberoamer. **1** (1985) 1–56.

[4] C. FEFFERMAN, *The Bergman kernel and biholomorphic mappings of pseudoconvex domains*, Invent. Math **26** (1974), 1–65.

[5] H. HEDENMALM, *The dual of a Bergman space on simply connected domains*, J. Anal. Math. **88** (2002), 311–335.

[6] N. KERZMAN and E. M. STEIN, *The Szegö kernel in terms of the Cauchy-Fantappié kernels*, Duke Math. J. **25** (1978), 197–224.

[7] L. LANZANI and E. M. STEIN, *Szegö and Bergman projections on non-smooth planar domains*, J. Geom. Anal. **14** (2004), 63–86.

[8] L. LANZANI and E. M. STEIN, *The Bergman projection in L^p for domains with minimal smoothness*, arXiv:1201.4148.

[9] M. RANGE, "Holomorphic Functions and Integral Representations in Several Complex Variables", Springer Verlag, Berlin, 1986.

Congruent numbers and Heegner points

Shou-Wu Zhang

1. Problem

An anonymous Arab manuscript[1], written before 972, contains the following

Congruent number problem (Original version). *Given an integer n, find a (rational) square γ^2 such that $\gamma^2 \pm n$ are both (rational) squares.*

Examples.

1. 24 is a congruent:

$$5^2 + 24 = 7^2, \qquad 5^2 - 24 = 1^2.$$

So is 6:

$$\left(\frac{5}{2}\right)^2 + 6 = \left(\frac{7}{2}\right)^2, \qquad \left(\frac{5}{2}\right)^2 - 6 = \left(\frac{1}{2}\right)^2.$$

It is clear that it suffices to assume n has no square factors.

2. Leonard Pissano in 1220's was challenged by Emperor's scholars to show that 5, 7 are congruent numbers:

$$5: \qquad \left(\frac{49}{12}\right)^2, \qquad \left(\frac{41}{12}\right)^2, \qquad \left(\frac{31}{12}\right)^2$$

$$7: \qquad \left(\frac{463}{120}\right)^2, \qquad \left(\frac{337}{120}\right)^2, \qquad \left(\frac{113}{120}\right)^2$$

Conjecture (Fibonacci). 1 *is not a congruent number.*

It took 400 hundreds year until it is proved by Fermat using his method of *infinite descent*.

---

[1] L. E. Dickson, *History of theory of numbers*, vol. 2 (1920), page 462.

Triangular version

Congruent number problem (Triangular version).   *Given a positive integer n, find a right angled triangle with rational sides and area n.*

This was considered as *a principle object of the theory of rational triangles in* 10*th century.*

The equivalence of two forms is not difficult to prove: Given arithmetic progression $\alpha^2, \beta^2, \gamma^2$ with common difference n then we have the following right triangle with area n:

$$a = \gamma - \alpha, \qquad b = \gamma + \alpha, \qquad c = 2\beta.$$

Conversely given a right triangle $[a, b, c]$ with area n, then we have following progression with difference n:

$$\left(\frac{a-b}{2}\right)^2, \qquad \left(\frac{c}{2}\right)^2, \qquad \left(\frac{a+b}{2}\right)^2.$$

The following are right triangles with areas 5, 6, 7:

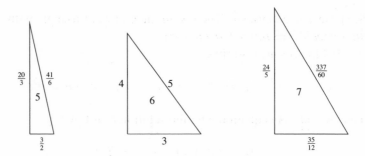

2. Fermat 1659

In a letter to his friend, Fermat wrote:

"I discovered at least a most singular method... which I call *the infinite descent.* At first I used it only to prove negative assertions such as... there is no right angled triangle in numbers whose area is a square, ... If the area of such a triangle were a square, then there would also be a smaller one with the same property, and so on, which is impossible, ..."

He adds that to explain how his method works would make his discourse too long, as the whole mystery of his method lay there. To quote Weil: "Fortunately, just for once he (Fermat) had found room for this mystery in the margin of the the very last proposition of Diophantus".

Fermat's argument was based on the ancient Euclidean formula (300 BC): Given (a, b, c) positive integers, pairwise coprime, and $a^2 + b^2 =$

c^2. Then there is a pair of coprime positive integers (p, q) with $p + q$ odd, such that

$$a = 2pq, \qquad b = p^2 - q^2, \qquad c = p^2 + q^2.$$

Thus we have a *Congruent number generating formula*:

$$n = pq(p + q)(p - q)/\square.$$

Here are examples:

$$
\begin{array}{lll}
(p, q) = (2, 1), & pq(p^2 - q^2) = 2 \times 3, & n(2, 1) = 6; \\
(p, q) = (5, 4), & pq(p^2 - q^2) = 5 \cdot 4 \cdot 9, & n(5, 4) = 5; \\
(p, q) = (16, 9), & pq(p^2 - q^2) = 16 \cdot 9 \cdot 7, & n(16, 9) = 7.
\end{array}
$$

Theorem (Fermat). $1, 2, 3$ *are non congruent.*

The following is the argument for non-congruent number 1:

1. Suppose 1 is congruent. Then is an integral right triangle with *minimum area*: $\square = pq(p + q)(p - q)$.
2. As all 4 factors are co-prime,

$$p = x^2, \quad q = y^2, \quad p + q = u^2, \quad p - q = v^2.$$

3. Thus we have an equation with the solution as follows:

$$(u + v)^2 + (u - v)^2 = (2x)^2.$$

4. Then $(u + v, u - v, 2x)$ forms a right triangle and with *smaller area* y^2. Contradiction!

3. Conjectures

Following Goldfeld and BSD, we have the following conjecture concerning the distribution of congruent numbers:

Conjecture. *Let n be a square free positive integer.*

1. *if $n \equiv 5, 6, 7 \mod 8$ then n is congruent.*
2. *if $n \equiv 1, 2, 3 \mod 8$ then n has probability 0 to be congruent:*

$$\lim_{X \to \infty} \frac{\#\{n \leq X : n = 1, 2, 3 \mod 8 \text{ and congruent}\}}{X} = 0.$$

Examples.

1. Congruent numbers under 23:

$$n = pq(p+q)(p-q)/\square.$$

$14 \equiv 6 \mod 8 \qquad (p,q) = (8,1);$
$15 \equiv 7 \mod 8 \qquad (p,q) = (4,1);$
$21 \equiv 5 \mod 8 \qquad (p,q) = (4,3);$
$22 \equiv 6 \mod 8 \qquad (p,q) = (50,49);$
$13 \equiv 5 \mod 8 \qquad (p,q) = (5^2 \cdot 13, 6^2);$
$23 \equiv 7 \mod 8 \qquad (p,q) = (156^2, 133^2).$

2. Conjecturally, if $n \equiv 1,2,3 \mod 8$ is congruent then *there are at least two very different ways* to construct triangles:

$34 \equiv 2 \mod 8, \qquad (p,q) = (17,1), \qquad (17,8);$
$41 \equiv 1 \mod 8, \qquad (p,q) = (25,16), \qquad (41,9);$
$219 \equiv 3 \mod 8, \qquad (p,q) = (73,48), \qquad (169,73).$

4. Theorems

The following are some results about the congruent and non-congruent numbers with specific prime factors.

Congruent primes.

Theorem (Genocchi (1874), Razar (1974)). *A prime p (resp. $2p$) is non-congruent if $p \equiv 3 \mod 8$ (resp. $p \equiv 5 \mod 8$).*

Theorem (Heegner (1952), Birch–Stephens (1975), Monsky (1990)). *A prime p (resp. $2p$) is congruent if $p \equiv 5,7 \mod 8$ (resp. $p \equiv 3 \mod 4$).*

Zagier have computed a precise triangle with prime area 157:

$$157 = \tfrac{1}{2}ab, \qquad a^2 + b^2 = c^2.$$

$$a = \frac{411340519227716149383203}{21666555693714761309610}$$

$$b = \frac{6803298487826435051217540}{411340519227716149383203}$$

$$c = \frac{224403517704336969924557513090674863160948472041}{8912333226892885958802553517896716357016480830}$$

Congruent numbers with many prime factors.

Theorem (Feng 1996, Li–Tian 2000, Zhao 2001). *For any positive integer k, and any $j \in \{1, 2, 3\}$, there are infinitely many non-congruent numbers n with k odd primes factors, and congruent to $j \mod 8$.*

Theorem (Gross 1985, Monsky 1990, Tian 2012[2]). *For any positive integer k, and any $j \in \{5, 6, 7\}$, there are infinitely many congruent numbers n with k odd primes factors, and congruent to $j \mod 8$.*

5. Elliptic curves

Congruent number problem (Elliptic curve version). *For a positive integer n, find a rational point (x, y) with non-zero y on the elliptic curve:*

$$E_n : \qquad ny^2 = x^3 - x.$$

The equivalence with triangler version is given by:

$$x = \frac{p}{q} \Leftrightarrow (a, b, c) = (2pq, \, p^2 - q^2, \, p^2 + q^2).$$

The rational points on an elliptic curves form a group. The understanding of this group structure is a major question in modern number theory and arithmetic algebraic geometry. The following was conjectured by Poincaré in 1901:

Theorem (Mordell 1922). *Let C be an elliptic curve over $\mathbb{Q}$. Then*

$$C(\mathbb{Q}) \simeq \mathbb{Z}^r \oplus C(\mathbb{Q})_{\text{tor}}.$$

for some $r > 0$, where $C(\mathbb{Q})_{\text{tor}}$ is a finite group.

L-series

Let Δ denote the discriminant of C and set

$$N_p = \#\{\text{solutions of } y^2 \equiv x^3 + ax + b \mod p\}.$$

$$a_p = p - N_p.$$

$$L(C, s) = \prod_{p \nmid 2\Delta} (1 - a_p p^{-s} + p^{1-2s})^{-1}.$$

Then $L(C, s)$ is absolutely convergent for $\Re(s) > 3/2$ (Hasse), and has holomorphic continuation to $\mathbb{C}$ (Wiles, *et al.*).

An $1,000,000 prize problem by Clay Math Institute:

Conjecture (Birch and Swinnerton-Dyer).   *The Taylor expansion of* $L(C, s)$ *at* $s = 1$ *has the form*

$$L(C, s) = c(s - 1)^r + higher\ order\ terms$$

with $c \neq 0$ *and* $r = \text{rank} C(\mathbb{Q})$. *In particular* $L(C, 1) = 0$ *if and only if* $C(\mathbb{Q})$ *is infinite.*

Application to congruent numbers

1. The L-series $L(E_n, s)$ has a functional equation $s \to 2 - s$ with sign

$$\epsilon(n) = \begin{cases} 1 & n \equiv 1, 2, 3 \mod 8 \\ -1 & n \equiv 5, 6, 7 \mod 8. \end{cases}$$

This gives a partition $\mathbb{N} = S \coprod T$ according to $\epsilon = \pm 1$.

2. Conjecturally, 100% of $n \in S$ are non-congruent numbers.
   This maybe be checked by computing the *Selmer groups* which is a modern version of the Fermat's infinite descent, *the only tool available for non-congruent numbers.*

3. Conjecturally, 100% of $n \in T$ are congruent numbers with solutions given by *Heegner points, the only tool available for congruent numbers.*

Tian's theorem

Theorem (Ye Tian). *Let* $m \equiv 5, 6, 7$ *be a square free number and consider* $E^{(m)} : my^2 = x^3 - x$. *Then*

$$\text{rank} E^{(m)}(\mathbb{Q}) = 1 = \text{ord}_{s=1} L(E^{(m)}, s)$$

provided the following condition verified:

1. *the odd part* $n = p_0 p_1 \ldots p_k$ *with* $p_i \equiv 1 \mod 8$ *for* $i > 1$.
2. *the class group* $\mathcal{A}$ *of* $K = \mathbb{Q}(\sqrt{-2n})$ *satisfies*

$$\dim_{\mathbb{F}_2}(\mathcal{A}[4]/\mathcal{A}[2]) = \begin{cases} 1, & p_0 \equiv \pm 1 \mod 8 \\ 0, & p_0 \equiv \pm 3 \mod 8 \end{cases}$$

6. Heegner method

Both Monsky and Tian proven their theorem based on the original method of Heegner. Heegner published his paper in 1952 as a 59 years old non-professional mathematician. In the same paper, Heegner solved Gauss class number one problem whose correctness is accepted by math community only in 1969, four years after Heegner died.

Modular parametrization

Heegner's main idea of constructing solutions to $E : y^2 = x^3 - x$ is by using modular functions (analogous to parametrizing the unit circle using *trigonometric functions* $(\cos 2\pi t, \sin 2\pi t)$):

$$f : \quad \mathcal{H} := \{z \in \mathbb{C}, \Im z > 0\} \longrightarrow E(\mathbb{C}).$$

The same idea can be used to answer the question:

Question. *Why is $e^{\pi\sqrt{163}}$ an almost an integer?*

$$e^{\pi\sqrt{163}} = 262537412640768743.99999999999925...$$
$$= 640320^3 + 744 - 74 \times 10^{-14}...$$

The answer lies in the algebraicity of the special values of modular functions just like trigonometric functions which are transcendental but take algebraic values at rational multiples of π. Modular functions are transcendental, but take algebraic values at quadratic points. For example:

$$j(z) = e^{-2\pi i z} + 744 + 196884 e^{2\pi i z} + 21493760 e^{4\pi i z} + \cdots$$

$$j((1 + \sqrt{-163})/2) = -640320^3 = -262537412640768000.$$

Heegner point

Here is a precise steps of construction of Heegener points in Tian's paper: Define $E^{(m)} : my^2 = x^2 - x, m^* := (-1)^{(n-1)/2} m$.

1. Take a standard parameterization $f : X_0(32) \to E^{(1)}$.
2. Take a CM point in $X_0(32) = \Gamma_0(32)\backslash\mathcal{H}$ by

$$P = \begin{cases} [i\sqrt{2n}/8], & n \equiv 5 \mod 8, \\ [(i\sqrt{2n}+2)/8] & n \equiv 6, 7 \mod 8 \end{cases}$$

3. Take $\chi : \mathrm{Gal}(H(i)/K) \longrightarrow \mathrm{Gal}(\sqrt{m^*})/K \simeq \{\pm 1\}$,
4. Define $P_m = \sum_{\sigma \in \mathrm{Gal}(H(i)/K)} f(P)^\sigma \chi(\sigma)$.

Then $P_m \in E(\mathbb{Q}(\sqrt{m^*})^- \simeq E^{(m)}(\mathbb{Q})$.

What Tian proves is the following non-vanishing statement of Heegner points:

Theorem (Ye Tian). *Assume the following condition verified:*

1. *the odd part $n = p_0 p_1 \ldots p_k$ with $p_i \equiv 1 \mod 8$ for $i > 1$.*
2. *the class group $\mathcal{A}$ of $K = \mathbb{Q}(\sqrt{-2n})$ satisfies*

$$\dim_{\mathbb{F}_2}(\mathcal{A}[4]/\mathcal{A}[2]) = \begin{cases} 1, & p_0 \equiv \pm 1 \mod 8 \\ 0, & p_0 \equiv \pm 3 \mod 8 \end{cases}$$

Then
$$P_m \in 2^k E^{(m)}(\mathbb{Q}), \qquad P_m \notin 2^{k+1} E^{(m)}(\mathbb{Q}).$$

COLLOQUIA

The volumes of this series reflect lectures held at the "Colloquio De Giorgi" which regularly takes place at the Scuola Normale Superiore in Pisa. The Colloquia address a general mathematical audience, particularly attracting advanced undergraduate and graduate students.

Published volumes

1. Colloquium De Giorgi 2006. ISBN 978-88-7642-212-6
2. Colloquium De Giorgi 2007 and 2008. ISBN 978-88-7642-344-4
3. Colloquium De Giorgi 2009. ISBN 978-88-7642-388-8, e-ISBN 978-88-7642-387-1
4. Colloquium De Giorgi 2010–2012. ISBN 978-88-7642-455-7, e-ISBN 978-88-7642-457-1

Fotocomposizione CompoMat srl, Loc. Braccone, 02040 Configni (RI)
Finito di stampare nel mese di maggio 2013
dalla CSR srl, Via di Pietralata 157, 00158 Roma